# CONVENTIONAL FLOWMETERS

*Conventional Flowmeters, Volume II,* covers the origin, principle of operation, development, advantages and disadvantages, applications, and frontiers of research for conventional flowmeters, which include differential pressure transmitters and primary elements, positive displacement, turbine, open channel, and variable area.

There are more conventional meters being used in the field than new-technology meters. New developments, such as more accurate pressure transmitters, new primary elements such as cone elements, reversible flow, and dual rotor turbine meters, and variable area meters with transmitters and signal outputs, are discussed.

**Features:**

- Offers a working knowledge of the origin and development of conventional flowmeters: differential pressure transmitters and primary elements, positive displacement, turbine, open channel, and variable area.
- Discusses the advantages and disadvantages of conventional meters and provides a rationale for retaining or replacing these meters.
- Focuses on the origin, development of operating principles, and applications for the meters.
- Explores the development of each conventional flowmeter type, including the roles of companies such as Siemens, ABB, Emerson, Foxboro, KROHNE, and Endress+Hauser.

This book is designed for anyone involved with flowmeters and instrumentation, including product and marketing managers, strategic planners, application engineers, and distributors.

# CONVENTIONAL FLOWMETERS

## Volume II

Jesse Yoder

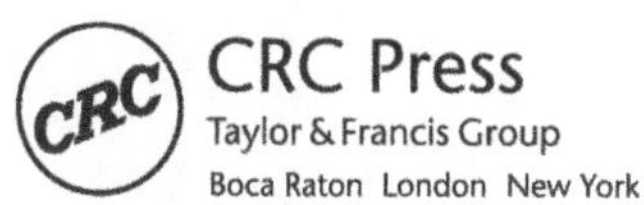


CRC Press is an imprint of the
Taylor & Francis Group, an **informa** business

First edition published 2023
by CRC Press
6000 Broken Sound Parkway NW, Suite 300, Boca Raton, FL 33487-2742

and by CRC Press
4 Park Square, Milton Park, Abingdon, Oxon, OX14 4RN

*CRC Press is an imprint of Taylor & Francis Group, LLC*



ISBN: 978-0-367-65543-3 (hbk)
ISBN: 978-1-032-35898-7 (pbk)
ISBN: 978-1-003-13002-4 (ebk)

DOI: 10.1201/9781003130024

Typeset in Times
by MPS Limited, Dehradun

# *Dedication*

*This book is dedicated to Belinda Burum, who has been a constant source of inspiration and insight to me for more than 32 years.*
*–Jesse Yoder*

# Contents

# List of Figures

# List of Tables

# Preface

I have developed the ideas in this book over the past 24 years of writing about flowmeters and related instrumentation. During this time, I have written more than 280 market studies and published close to 300 articles on flowmeters and instrumentation topics. Most of these articles reflected current research being done. I find that I often learn most about a topic by writing about it, and these articles have proved very instructive to me.

The main purpose of this book is to provide an in-depth look at the main types of flowmeters. This includes writing about their origin and historical development, along with their theory of operation. While these may seem like straightforward topics, they are not. There is sometimes disagreement about who first invented a technology, or what company first introduced a certain type of flowmeter. While I do not claim to have resolved all these issues in this book, my approach is to try to give my best judgment rather than just to present various alternatives.

My goal in describing the theory of operation of each flowmeter type is to present a clear explanation that is not overly technical. No doubt there are other writers who can present these topics in a more technical way with more equations. My goal is to explain in plain English how the different flowmeters work so the explanation is understandable to those who are not experts in the field. I use equations, photographs, and illustrations as necessary to achieve this goal. In some cases, I describe my own perspective while presenting the more traditional view so that the reader can make up his or her own mind.

This book is about conventional flowmeters. These include differential pressure (DP) flowmeters, which combine a pressure transmitter with a primary element that constricts the flow. It also includes positive displacement, turbine, open channel, and variable area flowmeters. While the markets for these flowmeters are not as fast-growing as those of new-technology flowmeters, they tend to have a larger installed base because they have been around longer than new-technology meters. They also have been studied extensively and are among the first to get industry approvals such as the approval for their use in custody transfer measurement.

The history of some of these flowmeters goes back to the 1790s. Tracing the development of these meters requires extensive research. In some cases, patent searches are required. Product managers and general managers at manufacturers of these flowmeters are an especially vital source of information, since they often know the history of the development of their own meters, and the developmental history of competitive meters. As part of my research into flowmeter development, I interviewed the founders of a number of flowmeter companies and asked them to explain how they came to start their companies. Some of these interviews are published at www.legendsofflow.com, and some are excerpted in this book.

When discussing the titles of these two volumes with the editors, I found that they prefer the term "conventional flowmeter" to "traditional technology flowmeter." In some ways I find this preferable because I always found "traditional technology flowmeter" to be somewhat wordy, and often substituted the term "traditional flowmeter" or "traditional meter" for it. For this reason, I was happy to go along with the proposed change in terminology. Therefore, the second volume in this set is called "Conventional Flowmeters." I have made a corresponding change in terminology at my company Flow Research, and our new studies now refer to conventional flowmeters instead of traditional technology flowmeters.

Most conventional flowmeters have moving parts, which makes them subject to more frequent calibration than new-technology meters. They also can wear out or go out of calibration more quickly than most new-technology meters. On the other hand, manufacturers are responding to this challenge by developing more durable parts and enhancing the diagnostic capabilities of conventional meters. For example, turbine meter manufacturers have introduced ceramic bearings, extensive diagnostics, redundant meters, and the capability of measuring reversible flow. All these innovations have made turbine meters more durable and appealing to end-users.

Chapter 11 includes a discussion of some fundamental concepts of geometry. Some of these ideas I developed while a student at Rockefeller University. Because most pipes are round, it is necessary to use the formula for the area of a circle in calculating pipe area. This involves the value $\pi$. I show graphically how the idea of $\pi$ arises by drawing a square inside a circle. The traditional formula for calculating the area of a circle involves determining how many squares fit inside a circle. Because circular and square area cannot be analyzed using the same unit of measure, no rational number will provide the correct answer. This is why mathematicians have resorted to $\pi$, an irrational nonrepeating value that stretches out to infinity.

Chapter 12 explores some of the relations between sensors and the mind. One of my main subjects in studying philosophy is to explain what the mind is, and how mind and body interact. In chapter 12 I treat the mind as a biological sensor. I think we can learn a lot about the mind by observing the analogies between biological and electronic sensors. I expect to explore this topic further in future writings, along with my views on circular geometry. Chapter 12 concludes with a discussion of a modified form of decimal time called flowtime.

My knowledge of flowmeters has come over 29 years from taking courses, reading books and articles, talking to thousands of professionals in the field, attending and speaking at conferences, and traveling to visit companies in the United States, Europe, the Middle East, and Australia. One thing I love about my job is that I learn something new nearly every day, and I never get bored. I have a dedicated staff at Flow Research who help me do the research and turn the raw data into a presentable study. I am one of those lucky people who loves what he does and sees each day as a new opportunity. I have tried to distill as much of this knowledge as possible into this book. I hope you enjoy it!

**Jesse Yoder, PhD**
*Flow Research, Inc.*

# Acknowledgments

I would first of all like to thank my philosophy professors who taught me how to write analytical philosophy. Throughout college I wrote dozens of philosophy papers, and when I went to graduate school at Rockefeller University I wrote three tutorial papers every week for several years. At the University of Massachusetts Amherst, I continued writing papers and served as a Teaching Assistant. I then wrote my PhD dissertation on philosophy of mind. The professors who had most influence over me at Rockefeller were Donald Davidson, Joel Feinberg, Harry Frankfurt, and Saul Kripke. At the University of Massachusetts Amherst, I am grateful to Gareth Matthews, who supervised my dissertation.

This book is about flow and philosophy. To me, they are very closely interrelated, The discussion of geometry in Chapter 11 and of sensors in Chapter 12 shows two ways they are closely connected. Many philosophical puzzles, including those about the number line, are related to the notion of continuity. Flow is a continuous phenomenon and understanding and analyzing continuity has become something of a life's work and goal for me.

This book is the second volume of the set called *Advances in Flowmeter Technology*. It centers on conventional flowmeters. I am especially grateful to the many people in the field who have supported my research, and the ongoing projects at Flow Research. They are too many to acknowledge them all, but I would especially mention Mark Heindselman, Scott Nelson, Bill Graber, Jacob Freeke, Mike Touzin, Karin Kettenbach, Dr. Jean-Philippe Herzog, Roberto Guazzoni, Steve Ifft, Bob Carroll, Eric Sanford, Jessica Lackey, Yuichi Kikuchi, Frank Frenzel, Robert Mapleston, Jim Peterson, Ken Ball, and Todd Sullivan. There are literally hundreds or perhaps I should say thousands of people who have helped me along the way since I began doing market research in 1991, and especially since founding Flow Research in 1998. Nick Limb helped me get started the first few years, and Brian Crosby has been a good friend since the beginning.

Belinda Burum has been a wonderful friend to me for 32 years and has served as a partner in building Flow Research. She has been my source of inspiration during this entire time. I could not have built Flow Research without her. In this book and in Volume I, she was especially helpful with the company profiles. Nicole Riordan, Gabriella DeCologero, Kaleigh Flaherty, and Leslie Buchanan have all been a joy to work with. I have worked for many years with the editors of *Process Instrumentation* (formerly *Flow Control*), *Fluid Handling*, and other publications who have helped me reach a broader audience with the results of the data collected by Flow Research. I am also grateful to CRC Press/Taylor & Francis Group for giving me the opportunity to write these two books.

Somehow my love for philosophy and flow have become joined, and I wake up every morning wanting to go into the office and work on whatever projects are going at the time. Just as much, I love to write, and I hope to continue writing for as long as God gives me the strength and capability.

More than anyone else I am grateful to Vicki Tuck for our many years of love, joy, and humor together. She has also contributed greatly to the success of Flow Research over the years.

**–Jesse Yoder**

# Author

**Jesse Yoder, Ph.D.** is president of Flow Research, Inc., a company he founded in 1998, which is located in Wakefield, MA. He has 31 years of experience as an analyst and writer in process control. He has authored more than 250 market research studies in industrial automation and process control and has written more than 280 published journal articles on instrumentation topics. He has published articles in *Flow Control, Fluid Handling, Processing, Pipeline & Gas Journal, InTech* magazine, *Control*, and other instrumentation publications.

Study topics include Coriolis, magnetic, ultrasonic, vortex, thermal, differential pressure, positive displacement, and turbine flowmeters. He has authored two separate six-volume series of studies on gas flow and oil flow. Dr. Yoder is a regular speaker at flowmeter conferences, both in the U.S. and abroad. Dr. Yoder studied philosophy at the University of Maryland, The Rockefeller University, and the University of Massachusetts Amherst, where he earned his Ph.D. in 1984. He served as an adjunct professor of philosophy for ten years at the University of Massachusetts Lowell and Lafayette College. In 1989 he co-founded the InterChange Technical Writing Conference, which he directed for six years.

In 2014 he founded the Flowmeter Recalibration Working Group (FRWG), dedicated to determining the criteria for deciding when a flowmeter should be recalibrated. The group is composed of 26 members that represent the leading flowmeter manufacturers and flow calibration facilities worldwide, along with end-users. The FRWG held its third meeting on June 21, 2018, and has just completed an extensive survey of flowmeter end-users. Details about the FRWG can be found at www.frwg.org.

Dr. Yoder has become a world-renowned authority and expert in the area of flow measurement and market research. As an entrepreneur, author, consultant, and inventor, he has helped define the concepts used in flow measurement, and is widely respected as an innovator in this field. He holds two U.S. patents for a new flowmeter design involving placing two tubes inside a pipe and putting a sensor on each type. The patents are called "Flowmeters for Large Diameter Pipes" and were granted in July 2015 and August 2017. Yoder owns 380 domain names, many of them related to flow and philosophy, and maintains over 200 active websites on matters relating to flowmeters and philosophical topics. His main website is www.flowresearch.com. Yoder lives in Wakefield, Massachusetts, where he enjoys racquetball, bird-watching, and writing philosophy.

# 1 A Preview of Coming Attractions

## OVERVIEW

This is a book about flow and flow measurement. Flow is all around us. Whether it is air flow, water flow, traffic flow, or the flow of gasoline into our cars, it is difficult to escape the impact of flow in our daily lives. Most of us take flow for granted, just as we take gravity and the presence of the sun and moon for granted. Yet, it would be difficult to live without flow. If you tried to imagine your life without all those things around us that flow, your life would be either bleak or nonexistent. It is impossible for humans to live without being able to breathe air, and life without water would be short-lived. So even if you have never been inclined to study flow, or even to wonder what it is, your life and the lives of billions of other people on this planet depend on it.

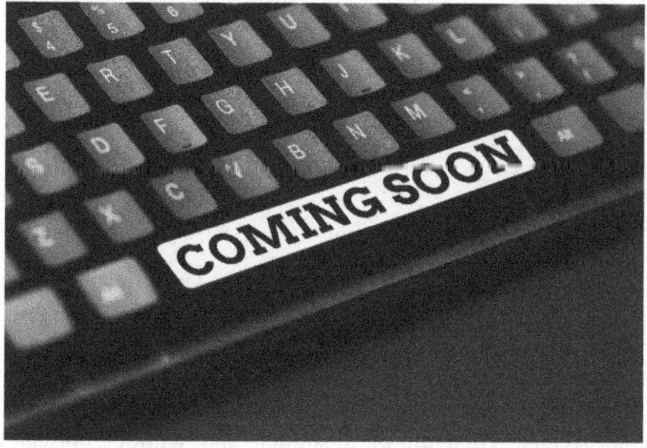


Measuring flow is another matter. Although we could probably survive without having to measure flow, there are many occasions when flow measurement is important. If you have ever baked a cake, you know the importance of accurate measurement of the liquids involved. You may use a measuring cup or a glass or a cup that is close to 8 ounces in size. When you fill up your tank with gas, you want to be assured that you are getting the exact amount of gasoline you are paying for. When buying prepackaged liquids such as milk, you probably assume that the level of the milk in the container accurately reflects the

DOI: 10.1201/9781003130024-1

quart or gallon you are buying. Most manufacturers do a good job with this. However, when I compared five of those familiar 500-millimeter water bottles (just over a pint), all marked 16.9 ounces, I was surprised to find that two of them had slightly lower water levels than the other three. This means that two had slightly less than 16.9 ounces unless the other three had slightly more. These differences are small enough that they generally go unnoticed, but they could reflect a slight inaccuracy in the bottling equipment. Almost every measurement is made with a certain tolerance for error.

The idea of how full a container is is actually more noticeable for solids than for liquids. Some cereal boxes and other containers with solid materials contain a disclaimer that reads "Sold by weight, not by volume." This is partly due to the fact that the contents may settle during shipping, and thus the container may not appear to be full. It may also be that the amount of material when weighed simply does not fill up the container, even before shipping. Solids such as grains and powders can flow, even though the volume that goes into a container may be determined by a scale rather than by a flowmeter. This book does not discuss the flow of solids, which are not fluids, but nonetheless, this is still an important part of the total flowmeter market.

Marketers play some interesting tricks when it comes to packaging products sold by weight in a container. Coffee is a good example. Coffee used to be sold in one-pound (16-ounce) cans. Then all of a sudden and without notice, the one-pound cans became 13 ounces. The cans looked the same as the one-pound cans, only they contained three ounces less of coffee. Apparently, this switch happened many years ago. There is an amusing story in the *Deseret News* with a date of May 1, 1989, by Mark Patinkin called "Investigating the Conspiracy: What Happened to the 1-Pound Can?" Mr. Patinkin's story recounts his calls to multiple coffee companies asking why the one-pound cans have been reduced to 13 ounces or even 11.5 ounces. The most prevalent answer he received was "Everyone else is doing it." In many cases, he was unable to reach anyone to give an answer.

Patinkin was able to reach several coffee companies, however. He was able to reach someone at Martinson, which at that time was still committed to the one-pound can. When asked about the future, an associate product manager said, "That decision is I suppose subject to business trends, I was told. If all the dinosaurs become extinct and you're a dinosaur … ." The idea is that if everyone else is selling the slimmed-down cans and you're still selling the one-pound can, you will lose the marketing battle and possibly the marketing war. Apparently, the pressure of the slimmed-down can eventually got to Martinson too, who decided they did not want to be among the dinosaurs. A quick check on Martinson coffee revealed the availability of a 10.3-ounce can of coffee in 2022.

Perhaps the most creative answer was given by someone named Joan at Folger's. Quoting again from the article:

> "Basically," explained Joan, "Folgers has developed a fast roast method that yields more from each bean. The grounds are more porous like gravel instead of denser like sand, allowing more water to come in contact with the surfaces."
>
> In other words, a new process puffs up the bean, giving it more surface, so when water hits it it yields more coffee. Thirteen ounces allegedly gives you the same bang that 16 used to.

The logic here is pretty obvious. If a company can sell less products for the same or similar price, and people don't notice the difference because the packaging remains the same, the company's profit margin goes up. Mark Patinkin concluded his article by predicting the 3.5 quart gallon of milk.

Those large vitamin bottles with many pills snuggled at the bottom, and the bulk of the bottle filled with cotton, raise a similar question. Vitamin D is a good example, because the pills are very small, yet they come in a container much larger than necessary to hold the pills. Here, though, the manufacturers have a better justification. They point out that often a larger bottle is necessary to print all the required text on the bottle. Also, the bottle size has to be large enough so that the "fine print" isn't too small. Another reason for the larger size is that it is more efficient and cost-effective to produce many bottles of a standard size than to produce bottles with a wide range of sizes. In addition, a bottle just large enough to hold 100 Vitamin D capsules would potentially get lost on the shelf alongside all the other larger bottles. So this provides more than a marketing justification for those larger vitamin bottles that hold very small pills, tablets, or capsules.

## CHAPTER SUMMARY

I would like to think that flow is such a fascinating topic that this book would be of interest to almost anyone. While this may be true, it is also true that the book is likely to be of most interest to those involved in some way with flowmeters, flow measurement, or who are working in industries in which flow measurement is important. The focus of the book is on the main types of conventional flowmeters. It includes a discussion of their origin, the history of their development including a focus on the companies involved, their principles of operation, and advances in the theory or production of these meters. The book also discusses reasons why people are buying these meters (growth factors), along with applications for them.

This book is the second of a two-part set called *Advances in Flowmeter Technology.* For classification and discussion purposes, it is convenient to divide flowmeters into two groups:

- *New-Technology Flowmeters*
- *Conventional Flowmeters*

This book, which is Volume II of this series, is called *Conventional Flowmeters.* Volume I is called *New-Technology Flowmeters*. The distinction between these two groups of meters is explained in Volume I, Chapter 2, although conventional flowmeters are described in Chapter 3 of this volume.

## A PREVIEW OF UPCOMING CHAPTERS

Chapter 2 begins by revisiting the definition of "flow." It examines definitions by leading writers on flow, including David Spitzer and Loy Upp. These definitions are compared to the one proposed in Volume I, Chapter 2. It makes the point that definitions are relative to the needs and applications of the definer, and so several different definitions of a term can be correct if they are given in different contexts. This does not mean that all definitions are equally correct, or that no definitions are wrong.

The next section discusses the cross-sectional area in light of the fundamental flow equation. The cross-sectional area of a round pipe is a circle formed when a plane intersects the pipe at a 90-degree angle. It points out that according to traditional mathematical theory a plane has only two dimensions, length and width, and so infinitely many of these circles can be formed by intersecting planes. However, if we say that points have area, lines have width, and planes have depth, then these circles will have three dimensions, and we can select the size of the cross-sectional area according to our measurement requirements.

The next section discusses velocity profile and Reynolds numbers. A Reynolds number is a dimensionless number that is the ratio of the inertial forces of flow to the viscous flow forces. In turbulent flow especially, the flow down the center of the pipe is faster than it is along the edges of the pipe. This has implications for insertion flowmeters, especially for ones that measure flow at only one point. The chapter discusses Pitot tubes and the development of averaging Pitot tubes. It concludes with a discussion of the multiple factors that influence flow measurement.

Chapter 3 defines conventional flowmeters. It then gives a brief overview of each type. Some of the topics discussed include history and origin, application, competitive position, and recent developments with respect to that meter type. The chapter continues with a discussion of why conventional meters have such staying power, and why we can expect to see them around for a long time. It concludes with a table of advantages and disadvantages of conventional flowmeters.

Chapters 4–9 focus on particular types of conventional flowmeters. These include the following:

- Differential Pressure (DP) Transmitters
- Primary Elements
- Positive Displacement (PD)
- Turbine
- Open Channel
- Variable Area

Generally speaking, each chapter has the following structure:

- Overview
- History and Development
- Advantages and Disadvantages
- Two Manufacturing Companies
- Types of Flowmeters (where relevant)
- Growth Factors
- Factors Limiting Growth
- Applications (where relevant)
- Frontiers of Research

Chapter 4 discusses the history and development of DP transmitters. DP transmitters are used with primary elements to create DP flowmeters. They have been around for over 100 years and have a large installed base. The chapter includes a table of advantages and disadvantages of DP transmitters and primary elements. It describes recent developments, including the introduction of multivariable DP transmitters. Companies described include Emerson Automation Solutions and

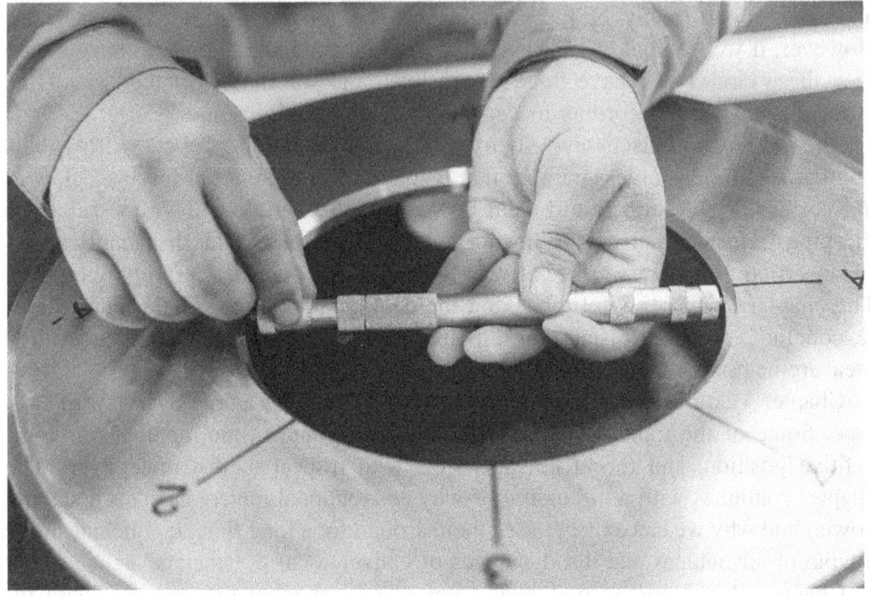

**FIGURE 1.1** Checking the opening in an orifice plate.

Yokogawa. It includes multiple growth factors for DP transmitters and concludes with a discussion of the frontiers of research for DP transmitters.

Chapter 5 describes many types of primary elements that are used with DP transmitters to create DP flowmeters (Figure 1.1).

These include the following:

- Orifice Measuring Points
- Pitot Tubes
- Venturis
- Cone Elements
- Flow Nozzles
- Wedge Elements
- Combination Elements (e.g., Nozzle/Pitot Tubes)
- Others (e.g., Dall Tubes, Laminar Flow Elements)

Chapter 5 includes pictures of the different types of primary elements. Daniel and McCrometer are the featured companies. The chapter continues with a discussion of growth factors and applications. It concludes by identifying frontiers of research for primary elements.

Chapter 6 discusses many types of PD flowmeters and includes illustrations showing how they work. PD meters are widely used downstream from oil refineries to measure the distribution of refined fuels. The following types of PD meters are included in this chapter:

- Oval Gear
- Rotary
- Gear
- Helical
- Nutating Disc
- Piston
- Diaphragm
- Spur Gear

Chapter 6 includes illustrations of the operating principles of the different types of PD meters. Featured PD flowmeter suppliers in Chapter 6 include Dresser Utility Solutions and TechnipFMC. The chapter continues with a discussion of growth factors and concludes by identifying frontiers of research for PD meters.

Chapter 7 discusses the history and development of turbine flowmeters. Turbine flowmeters were invented in the 19th century and were first used industrially to measure the use of fuel by military aircraft in World War II. Today they are widely used for custody transfer of natural gas.

The following are the main types of turbine flowmeters:

- Axial
- Paddlewheel
- Pelton Wheel
- Propeller
- Single Jet
- Multi-Jet
- Woltman

Honeywell Elster and Faure Herman are the two major turbine meter suppliers described in this chapter. Both growth factors and limiting factors for turbine meters are discussed. Another section looks at applications for turbine meters. The chapter concludes with frontiers of research for turbine flowmeters.

Chapter 8 describes open channel flow. It discusses the amount of water in the world and the rising importance of measuring water. This chapter compares open channel flow to closed-pipe flow and explains why there are more flowmeters devoted to closed-pipe flow than to open channel flow. The two main types of open channel flow measurement include

- Hydraulic Structure – Weirs and Flumes
- Area Velocity

In addition to these two main types, there are three other less commonly used methods of open channel measurement:

- Dilution
- Timed-Gravimetric
- Manning Formula

These are also described in Chapter 8. The open channel companies described in this chapter include Hach and Siemens. The chapter includes both growth and limiting factors for open channel meters. It concludes with a discussion of the frontiers of research for open channel flowmeters.

Chapter 9 describes variable area flowmeters, including their history and development. While these are the least expensive and least accurate type of flowmeter, they still play an important role in today's flowmeter world. One advantage they have is that they can operate without power. And in situations where end-users simply want a flow/no-flow indication, they are an ideal solution.

ABB and Brooks Instrument are the two companies highlighted in this chapter. The chapter also discusses both growth factors and limiting factors for variable area meters. It includes a short section on applications, along with a discussion of the frontiers of research for variable area meters.

Chapter 10 focuses on the oil and gas industry. This is the largest process industry that flowmeters are sold into. In addition, all the other process industries rely on the oil and gas industry for fuel. The flowmeter markets have been severely affected by downturns in the oil and gas industry in 2016 and in 2020. They were especially affected by lower crude oil prices that greatly reduced upstream exploration and production activities. In 2020, demand for refined fuels was also reduced.

Chapter 10 begins by distinguishing five types of fluids:

- Petroleum Liquids
- Non-petroleum Liquids
- Gases
- Industrial Gases
- Natural Gas

It then describes discusses oil prices from 2014 to 2018. A large part of the reason these prices have varied is due to the Organization of Petroleum Exporting Countries (OPEC). The chapter describes the founding of OPEC in 1960 and reveals that it was founded in response to a Western-style cartel consisting of seven major oil producers. These producers were charging higher prices to Middle Eastern countries than to Canada and Latin America.

The rest of the chapter looks at oil prices in 2018 and 2019, and the effect of the COVID-19 pandemic on oil prices and the flowmeter market. It discusses the recovery of oil prices in 2021 and also the high oil prices in 2022. It concludes with a discussion of the effect of these prices on the flowmeter market.

Chapter 11 looks at the geometry of flow. Because most flow that is measured occurs in pipes, and most pipes are round, we need to use the formula for the area of a circle to compute the area of pipes. Knowing this area is essential to knowing the volume of flow passing through a pipe. According to Euclidean geometry, the area of a circle is $\pi r^2$. This is equivalent to saying how many squares fit into a circle. There is no rational value for this number. The only way to give a rational value for the area of a circle is to start with a different unit of measurement such as a round inch.

Chapter 11 also discusses the way in which calculus computes the area under a curve. It does so by assuming that this area can be conceived of as the limit of infinitely many rectangles. I argue that this is an implausible solution to the problem of area under a curve because an infinite series can never be completed and a rectangle with no area is not a rectangle at all.

Chapter 12, Sensing and Measuring, defines what a sensor is. It then discusses the role of transducers and transmitters in flow measurement. The chapter describes three types of sensors:

- Mechanical
- Electronic
- Biological

The biological section describes the analogy between the human mind and electronic sensors.

The final section discusses whether clocks sense time. It continues with a discussion of decimal time. The chapter concludes by proposing a new form of time called flowtime. Flowtime retains some of the elements of decimal time but is more intuitive than decimal time.

# 2 The Building Blocks of Flow

## OVERVIEW

In Volume I of this two-volume set, Chapter 2 begins with a discussion of the question "What is flow?" This section talks about common examples of flow, including water flow, river flow, time flow, traffic flow, and other similar examples. It also describes a situation where something like flow occurs, but we don't call it flow. For example, a baseball hit "out of the park" doesn't flow out of the park; it flies out of the park. Likewise, a runner may be engaged in continuous motion while running a marathon, but we don't describe him or her as "flowing."

After some additional discussion in Chapter 2, I proposed the following definition of flow:

> **Flow is the continuous and uninterrupted motion of a fluid or a pattern of objects moving uniformly along a path in a direction.**

## DEFINITIONS ARE RELATIVE

Providing this definition of "flow" reminded me of an email conversation I had recently with the ISA (International Society of Automation) Director of Standards. I used to serve on the ISA SP50 Committee on Instrumentation Terminology. Our charter was to arrive at a group of definitions for widely used terms relating to instrumentation. While we had a group of definitions to work from that clearly needed to be updated, the Committee struggled to find a consensus on how to proceed in making this update. In the end, the Committee was either disbanded or simply failed to continue to meet.

I recently inquired about the status of this Instrumentation Terminology Committee. I received this response from the ISA Director of Standards, Charley Robinson:

> A big issue with terminology is that many people and organizations have different definitions for the same term, based on their needs and applications. Getting agreement is a real challenge.
>
> I typically refer people to the online IEC Electropedia.

Charley Robinson makes a very good point here. He is right that the definition of a term can vary depending on the needs and applications of the definer. Even so, there are dictionaries and glossaries, and people rely on a common understanding of the terminology in order to communicate. Dictionaries deal with the differences in

DOI: 10.1201/9781003130024-2

meaning by providing multiple definitions, but often it is easy to figure out which definition applies to one's own interests. It is a common practice in books of this type to publish a glossary of terms to give readers a quick understanding of relevant terminology used in the book.

My response to Charley's comment was

> Maybe what's needed then is for someone to collect the existing glossaries to see how they compare. Some companies have done their own as well. If you're right, then what terminology you use is relative to your situation and needs, and there is not always a "right" answer.

I think that creating a collection of glossaries of instrumentation terms would be useful. However, the questions would still remain about which ones are most relevant, and is it possible to consolidate them into a "master glossary" that would recognize that definitions may differ depending on context? I am still a passionate believer in the need for an agreed-upon glossary of instrumentation terms, whether that is done by collecting existing glossaries or by formulating an updated version of one or more glossaries.

In light of this discussion, it is worthwhile taking a look at how some of the leading authors in the field have defined "flow." Although this bears out Charley Robinson's point that definitions are relative to the needs and applications of the person making the definition, it also shows that there are some commonalities to these definitions that can provide for a single definition, or set of definitions, that the people proposing these definitions can agree on.

## DAVID SPITZER'S DEFINITION OF "FLOW"

A good place to start is with David Spitzer's excellent book, *Industrial Flow Measurement*. Spitzer defines "flow" as follows:

> Flow can be defined as the actual volume of fluid that passes a given point in a pipe per unit time. This can be expressed as:

$$Q = A \times v$$

We can immediately see how Spitzer's straightforward definition relates to the needs of his book. *Industrial Flow Measurement* is a comprehensive look at the different ways that flow is measured in an industrial context. He describes the different types of flowmeters, has multiple equations and illustrations for them, and also discusses important flow-related concepts such as viscosity, turndown, and many other ideas that are important for understanding flow measurement. While he briefly describes open channel flow, nearly the entire book is dedicated to describing fluid flowing through a pipe. As a result, he is not trying to give a definition of "flow" that includes traffic flow, the flow of time, or even the flow of solids such as grains or powders. Yet, Spitzer's definition makes perfect sense for his specific objectives.

## LOY UPP'S DEFINITION OF "FLOW"

Loy Upp in his outstanding book *Fluid Flow Measurement* takes a broader approach to defining "flow" by starting with some dictionary definitions and then putting them together into a definition that combines them into a single idea:

> The subjects below form the background for fluid flow measurement and that should be understood before embarking on the task of choosing a flow measurement system. "Fluid," "flow," and "measurement" are defined in generally accepted terms (Webster's New Collegiate Dictionary) as:
>
> **Fluid** – a. Having particles that easily move and change their relative position without separation of the mass and that easily yields to pressure; b. A substance (as a liquid or a gas) tending to flow or conform to the outline of its container.
>
> **Flow** – a. To issue or move in a stream; b. To move with a continued change of place among the consistent particles; c. To proceed smoothly and readily; d. To have a smooth uninterrupted continuity.
>
> **Measurement** – a. The act or process of measuring; b. A figure, extent, or amount obtained by measuring.

Combining these into one definition for fluid flow measurement yields

> **Fluid Flow Measurement** – The measurement of smoothly moving particles that fill and conform to the piping in an uninterrupted stream to determine the amount flowing.

Upp does not explain his use of the term "particles," which he takes from the dictionary definition, but it could be that he would appeal to molecular theory according to which fluids are composed of molecules and atoms. Alternatively, he could take sense (b) of the definition that uses the phrase "substance (as a liquid or a gas)" instead of sense (a), which refers to particles. This would yield the following definition:

> **Fluid Flow Measurement** – The measurement of a liquid or gas that fills and conforms to the piping in an uninterrupted stream, to determine the amount flowing.

## A COMPARISON OF DEFINITIONS

I will repeat my definition of "flow" from Volume I, Chapter 2 in light of the above discussion:

> **Flow is the continuous and uninterrupted motion of a fluid or a pattern of objects moving uniformly along a path in a direction.**

Spitzer's definition includes the term "fluid," while Upp's modified definition includes "liquid or gas," which is equivalent to fluid. My definition also includes the term "fluid." Both my definition and Upp's definition include the idea that flow is "uninterrupted," while Spitzer does not specify that flow is uninterrupted or continuous. The mention of "pattern of objects" in my definition is somewhat similar to Upp's original definition that includes "smoothly moving particles." Both Spitzer and Upp mention flow being in a pipe, or piping, while I use the broader term "path." A path can include a pipe, but clearly David Spitzer and Loy Upp are writing a definition more specifically for an industrial process environment that involves measuring flow in closed pipes. My definition is broader in an attempt to include phenomena such as traffic flow and crowd flow, but it can be made more specific by substituting "pipe" for "path."

The above discussion shows that, while definitions can be right or wrong, the mere fact that two definitions are different doesn't show that one is right or one is wrong. Each one may be correct when relativized to the purpose of the definition and the needs of the definer. I would say that of the three definitions considered, Spitzer's is the narrowest of the three and mine is the broadest. But Spitzer's definition is designed specifically for closed-pipe flow, whereas mine attempts to include broader concepts such as traffic flow. Furthermore, there are common elements to all the above definitions that any correct definition of flow needs to include.

## CROSS-SECTIONAL AREA OF A PIPE

Chapter 2 of Volume I gives the fundamental flow equation:

$$Q = A \times v$$

Here $A$ is the cross-sectional area of the pipe at a specific point. This cross-sectional area is defined as the area of a circle that is bounded by the inside diameter of the pipe. In Volume I, Chapter 2, this area is defined in terms of the formula for the area of a circle:

$$\pi r^2$$

The cross-sectional area of a pipe is the portion of the pipe that coincides with an imaginary plane that slices through the pipe. While a plane can slice through a pipe at any angle, for flow measurement purposes it is generally considered as slicing through at a 90-degree angle, or perpendicular to the axis of the pipe.

When a plane slices through a pipe at this angle, it forms a circle in the area of intersection. However, when it slices through the pipe at different angles, it can form other geometric shapes. In general, a geometric plane can slice through any three-dimensional object and form different geometric shapes in doing so.

Planes are considered as being two-dimensional, meaning they have only length and width. A plane that cuts through a three-dimensional pipe is conceived of as creating a two-dimensional circle inside the pipe. The circumference of this circle coincides with the inner pipe wall. Like the plane, the circle has two dimensions: length and width, which in this case are the same. This is a necessary consequence of the conception of a plane as having only two dimensions, and as not having height or depth. It is also a consequence of considering a line as being one-dimensional, and of only having the attribute of length. This is discussed in Volume I, Chapter 10.

The same issue arises for defining the cross-sectional area that arises from defining points as having no area. How many two-dimensional circles can a two-dimensional plane create or inscribe? The answer has to be infinitely many because the circumference of the circles does not have width. Of course, we don't need infinitely many circles to measure flow in a pipe.

## AN ALTERNATIVE VIEW OF THE WIDTH OF CIRCLES

Everything stems from the definition of a point. If a point has no area, and a line is the path of a moving point, then the line has no area or width. Likewise, if we conceive of a plane as the path of a line in motion, then the plane will not have height or depth and will have only the two dimensions of length and width. The result is that a two-dimensional plane that intersects a pipe creates or inscribes a two-dimensional circle.

If instead, we model our number system on time, which has duration, we get a different result. Like the number line, units of time can be indefinitely small. For example, a nanosecond is one billionth of a second. However, unlike the idea of a point, there is no such thing as a unit of time with no duration. Nothing that exists can exist for zero seconds; saying that something exists for zero seconds is like saying it didn't happen at all. Even an explosion that seems to occur "instantaneously" takes place over a very brief period of time. What we do with time is to define the smallest unit that is convenient to deal with (e.g., the second, or the hour) and describe how long it takes for events to occur in terms of those small units. Seconds and minutes do petty well for daily life, whereas some computers require smaller units like nanoseconds.

If we use this approach for measuring flow through a pipe, then the cross-sectional area will have the third dimension of width, however small, and we can measure how quickly this three-dimensional area moves through the pipe. We can actually keep Spitzer's definition of "flow" as "the actual volume of fluid that passes a given point in a pipe per unit time." However, instead of being a dimensionless point, the point will be three-dimensional and of whatever size is most convenient for the measurement being made.

Of course, this still leaves us with using the irrational number $\pi$ to determine the area of the cross-sectional circular area. Just as the area of a circle is defined in terms of how many squares can fit into it, so flow through a pipe is defined in terms of cubic feet or cubic meters. The number $\pi$ will be with us until someone figures out a way to reconcile square area and round area, or we adopt some form of circular geometry.

## VELOCITY PROFILE

Flow within a pipe is not uniform. Flow tends to be fastest in the center of the pipe, and slower along the edges next to the pipe wall. This phenomenon is referred to as velocity profile, and it depends on a combination of inertial forces and viscous forces. The inertial forces act to push the flow along, whereas the viscous forces tend to slow it down.

The relationship between inertial and viscous forces was studied by Osborne Reynolds, an Irish-born innovator in the study of fluid mechanics and heat transfer. In 1883, he performed a famous experiment in which he squirted a jet of water containing dye into water flowing through a glass pipe. With the aid of a control valve, he was able to see both laminar (smooth) and turbulent flow. As a result of his experiments, he proposed a relationship between inertial and viscous forces called the Reynolds number:

$$Re = \frac{\text{Inertial Forces}}{\text{Viscous Forces}}$$

This is a dimensionless number, and it is quantified as follows:

| | |
|---|---|
| Laminar | <2,000 |
| Transitional | 2,000–4,000 |
| Turbulent | >4,000 |

For more details on the above equation, and the variables involved, refer to David Spitzer's book *Industrial Flow Measurement.*

## INSERTION FLOWMETERS

Velocity profile has important implications for flow measurement. Some flowmeters only measure flow at a point and then infer the total amount of flow going through the pipe from that one measurement. This holds true for many insertion meters. Insertion meters have a major advantage when measuring flow in very large pipes, such as those above 20 inches in diameter. By inserting a flowmeter into the pipe, they can determine flow velocity at a point and make a flowrate calculation based on that one measurement. The upside of insertion flowmeters is cost – they are less expensive than flanged or wafer meters because they avoid the cost of the meter body. The downside is accuracy – they may not be able to accurately calculate pipe flow based on measuring flow at one point.

Some popular insertion meters include magnetic, ultrasonic, turbine, vortex, and thermal. Knowledge of flow profiles becomes important for insertion meters. If the flow is laminar, a single-point measurement has a better chance of being accurate than if it is turbulent. In turbulent flow, the fluid in the pipe is moving faster in the center of the flowstream than at the edges. It is important to know where in the flowstream the measurement is being made. Insertion meters are likely to register a higher flowrate if the measurement is made in the center of the pipe than if the measurement is made at the pipe edge.

An example of how this fact has influenced the development of flow is the Pitot tube. Pitot tubes were invented by a French engineer named Henri Pitot in the early 18th century. They are mainly used to measure air flow, although they can also be used for liquids. Initially, Pitot tubes measured flow at a single point. Later, averaging Pitot tubes were developed with multiple ports that measured flow at multiple points in the flowstream. These averaging Pitot tubes were more accurate than the single-point variety. Pitot tubes are a type of primary element and base their flow calculation on the difference in pressure between the impact pressure on the front side of the Pitot tube and the lower pressure that develops on the backside. A pressure transmitter is used to compute flow, using Bernoulli's theorem.

## MULTIPLE FACTORS INFLUENCE FLOW MEASUREMENT

At the beginning of this chapter, we looked at several definitions of flow that appear to define flow as occurring in closed pipes. In reality, there is another type of flow called open channel flow that occurs in rivers, streams, wastewater treatment plants, and anywhere flow occurs that is not under pressure. These open channel meters are discussed in Chapter 8 of this volume. Flow through partially filled pipes is considered open channel flow. Open channel flow occurs when fluid is flowing but is not under pressure. Typically, this is gravitational flow.

Some of the main factors that influence flow include viscosity, the amount of pressure in the pipe, the amount of upstream and downstream straight flow, and the presence of elbows before the flowmeter. Some flowmeters work on almost any type of fluid, whereas others require that the type of flow should be programmed into the meter. Thermal flowmeters typically need to know what type of gas they are measuring. Ultrasonic flowmeters for liquids work differently than ultrasonic flowmeters for gas because the speed of sound is different for these two types of fluids. Magnetic flowmeters can only measure conductive liquids and they cannot measure steam or gas flow at all.

Given the complexities of flow measurement and even the differences in defining what flow is, it is no wonder that there are many different technologies for measuring flow. This volume describes conventional meters: their operating principles, some of the main companies that make them, and growth factors for these meters. It also describes the frontiers of research for these meters. When reading through these chapters, keep in mind this principle, "If it were too easy, it wouldn't be interesting." Like the medium itself, flow measurement is always changing, and adapting to new challenges. I have devoted a good part of my life to studying flow measurement, and I still learn something new nearly every day. I hope that some of the excitement of this enterprise comes across in this book and in Volume I.

# 3 Conventional Flowmeters

## OVERVIEW

Despite the growth of new-technology flowmeters such as Coriolis and ultrasonic over the past few years, conventional flowmeters continue to hold their own. Many users are still selecting differential pressure (DP), positive displacement (PD), turbine, variable area, and open channel as their flowmeter solutions. This chapter describes some of the more important recent developments for conventional flowmeters.

## DEFINING CONVENTIONAL FLOWMETERS

Conventional flowmeters share the following characteristics:

1. As a group, these meters were introduced before 1950.
2. They are less focused on new product development than new-technology meters.
3. Their performance overall – including criteria such as accuracy – is not similar to new-technology flowmeters.
4. They generally have higher maintenance requirements compared to new-technology flowmeters.
5. They are slower to incorporate recent advances in communication protocols and networking schemes such as HART, Foundation™ Fieldbus, and Profibus.

Conventional flowmeters include measurement technologies that have been around for decades and, in some forms, for centuries. Business is strong with many of these stalwarts in the industry. Why are customers still choosing them as preferable to newer alternatives?

## DIFFERENTIAL PRESSURE TRANSMITTERS

DP transmitters have been in use for more than a century to measure flow. DP transmitters rely on a constriction, called a primary element in the flowstream, to create a pressure drop in the line. They measure the difference between the downstream and the upstream pressure to compute flow, using Bernoulli's theorem. DP transmitters rely on a variety of primary elements to create the constriction in the line.

DOI: 10.1201/9781003130024-3

DP transmitters themselves are part of a family of pressure products. Other types of pressure transmitters include absolute, gauge, and multivariable. Absolute pressure transmitters measure pressure without taking atmospheric pressure into account. Gauge pressure transmitters include atmospheric pressure in their pressure measurement. Between these two, gauge pressure transmitters are more common than the absolute variety.

One of the most important developments in pressure transmitters with implications for flow has been the development of multivariable pressure transmitters. Multivariable transmitters measure two or more process variables. In the context of flow, their differential pressure capability enables them to compute volumetric flow. In addition to this measurement, they contain a pressure and temperature sensor and/or transmitter to measure temperature and/or pressure. Multivariable pressure transmitters that measure differential pressure, temperature, and gauge or absolute pressure can compute mass flow. This makes them capable of measuring the mass flow of gas and the more difficult steam flows.

## PRIMARY ELEMENTS

While there are many types of primary elements, orifice plates are the most common type. Orifice plates usually consist of a round metal (typically steel) plate with a hole or "orifice" in it. The purpose of the hole in the plate is to force fluid in a pipe to pass through a measurement point of a smaller diameter, thereby creating a pressure drop downstream of the plate. There are many types of orifice plates with different shape openings positioned at different locations on the plate. Most common variations include concentric, eccentric, and segmental.

Experiments with DP transmitters using orifice plates to measure gas flow took place in the late 1920s, culminating with the publication of AGA-1 in 1930. This was a report by the American Gas Association (AGA) that constituted the first "industry standard" for orifice plate meters. Further testing resulted in the publication of AGA-2 in 1935 and AGA-3 in 1955. These reports gave orifice plate flowmeters a dramatic lead in establishing an installed base among flowmeters for custody transfer applications. It wasn't until 1981 that the AGA published a similar report for turbine meters, called AGA-7. And in 1998, the AGA published AGA-9, a report on the use of ultrasonic flowmeters for custody transfer of natural gas.

Other types of primary elements include Venturi tubes, flow nozzles, Pitot tubes, wedge elements, and laminar flow elements. Of these types, Venturi tubes, flow nozzles, and Pitot tubes are the most common. Venturi tubes bend up at one end and are elongated at the other end. They are often used for large pipe applications, including water and wastewater. Flow nozzles are often used to measure steam flow. The most popular form of Pitot tubes is averaging Pitot tubes. These are used to measure air flow and other forms of gas flow. They are also commonly used to measure stack and exhaust gas emissions.

## POSITIVE DISPLACEMENT

The history of PD flowmeters goes back to 1815 when Samuel Clegg invented the first PD gas meter. This first meter was a water-sealed rotating drum meter. In 1843, Thomas Glover invented the first "dry" PD meter. Glover's meter contained a sliding valve and two diaphragms.

Today's diaphragm meters are similar in concept but use more modern materials. They remain distinctive, however, in being the only flow measurement technology that directly measures the volume of fluid by using precisely identical rotating compartments that capture a prescribed amount of fluid that passes through the meter, and then uses the rate of rotation caused by the fluid flow to determine the volume of fluid passing per time interval. The rotor's rotational velocity is directly proportional to the flowrate causing the rotation. Bopp & Reuther of Germany was the first to patent oval gear meters – a popular design version today – in 1932.

Nowadays, one of the main uses of PD meters is in gas utility billing applications. There are two main types of PD meters used for this purpose: diaphragm meters and rotary meters. Rotary meters are replacing diaphragm meters in many cases. Rotary meters are smaller and lighter than diaphragm meters. This replacement is occurring for other gas applications as well. Honeywell Elster is a dominant supplier of these meters in this market.

Oval gear meters are quite popular for oil applications, especially for downstream oil distribution involving custody transfer. Here they compete with Coriolis meters, which are gaining market share in downstream oil measurement due to their perceived accuracy and low maintenance.

PD meters do best in line sizes between 1.5 and 10 inches. It is unusual to find PD meters in line sizes above 10 inches. High accuracy is one strength of PD meters. They are also very good for measuring fluid with low flowrates. Downsides of PD meters include causing pressure drop and being essentially a mechanical meter with moving parts that are subject to wear.

## TURBINE

The first turbine flowmeter was invented by Reinhard Woltman in 1790. This makes the turbine the earliest meter invented among all of the flowmeters in use today. Turbine meters precede DP flowmeters by at least 100 years. Despite their early invention, it wasn't until after the World War II that they started making an impact on industrial markets. During the World War II, they were used to measure fuel consumption on military aircraft. Soon after this period, they were used in the petroleum industry to measure the flow of hydrocarbons.

In 1953, turbine meters were used to measure gas flow. Rockwell introduced turbine meters to the gas industry in 1963. Within 10 years, they were widely used in the gas industry for measuring gas flow. In 1981, the AGA published its report AGA-7, "Measurement of Fuel Gas by Turbine Meters." Since then, turbine meters have become firmly entrenched in the gas industry, especially for custody transfer applications.

Turbine meters actively compete with ultrasonic and DP flowmeters for measuring custody transfer of natural gas. They are widely used for custody transfer of natural gas in large natural gas pipelines. Turbine meters remain a viable choice for measuring steady, medium- to high-speed flows. They are more complementary than competing with PD meters because they perform well in larger line sizes (above 10 inches). The drawbacks of turbine meters are that they cause pressure drop and have moving parts (mainly the rotors) that wear.

## OPEN CHANNEL

Open channel flow occurs when liquid flows in a conduit or channel with a free surface. Rivers, streams, canals, and irrigation ditches provide examples of open channel flow. What is slightly confusing about this terminology is that the flow of liquids in partially filled pipes, when not under pressure, is also considered open channel flow. For example, water flowing through a culvert running underneath a street is considered open channel flow. Likewise, flows in sewers and tunnels are classified as open channel flows, along with other closed channels that flow through partially filled pipes. Other examples of open channel flow include flow in water treatment plants, storm and municipal sewer systems, industrial waste applications, sewage treatment plants, and irrigation systems.

**Use of Weirs and Flumes:** A very common method of open-channel flow involves the use of a hydraulic structure such as a weir or flume. These hydraulic structures are called primary devices. A primary device is a restriction placed in an open channel with a known depth-to-flow relationship. Once a weir or flume is installed, the measurement of the depth of the water at a certain point is used to calculate flowrate. Charts are available that correlate various water depths with flowrates, taking into account different types and sizes of weirs and flumes.

**Area Velocity:** Flow can be measured without a hydraulic structure such as a weir or flume. In the area–velocity method, the mean velocity of the flow is calculated at a cross-section, and this value is multiplied by the flow area. Normally, this method requires two measurements: mean velocity and depth of flow.

Flowrate $Q$ is determined according to the continuity equation:

$$Q = A \times v$$

The area–velocity method is used when it is not practical to use a weir or flume, and for temporary flow measurements. Examples include influx and infiltration studies and sewer flow monitoring.

## VARIABLE AREA

Most variable area (VA) flowmeters consist of a tapered tube that contains a float. The upward force of the fluid is counterbalanced by the force of gravity. The point at which the float stays constant indicates the volumetric flowrate, which can be often read on a scale on the meter tube. VA meter tubes are made of metal, glass, or plastic. Metal tubes are the most expensive type, whereas plastic tubes are the least expensive. Metal tubes are used for high-pressure applications.

While most VA meters can be read manually, some also contain transmitters that generate an output signal for a controller or recorder. Although VA meters should not be selected for highly accurate requirements, they perform very well when a visual indication of flow is required. They are very effective at measuring low flowrates and can also serve as flow/no-flow indicators. VA meters do not require electric power, and thus can be safely used in flammable environments.

One important development for variable area flowmeters is the addition of transmitter output to meters. The Highway Addressable Remote Transducer (HART) protocol is available on some meters. This turns the VA meter into more than a visual indicator for recording the measurement and allows it to do control and recording. A class of VA meters called purgemeters has been developed to handle a variety of low-flow applications. Other areas of research include float design and materials of tube construction, especially metal.

## FAMILIARITY BREEDS RESPECT

While the explanations vary with the type of meter, there are several themes that run throughout. One answer is **familiarity**. End-users prefer using a technology they are familiar with and can understand. DP, PD, and turbine meters, in particular, are very well-known and well-understood technologies. Users have a comfort level with these technologies that are less likely to exist with the newer technologies such as Coriolis and vortex. In case more meters need to be added to a plant, users often stay with what they have rather than selecting a different type of meter.

A second reason is the **installed base**. Some flowmeters such as DP and PD have been around for over 100 years. Once these meters are installed, customers find in many cases that it is easier to replace them with similar meters than to switch to another technology as long as the demands of the application will still be met satisfactorily. Once a technology is in place, backup parts are readily available, any potential problems are usually known, and the path for replacement is clear. All these are reasons to stick with an existing technology.

Another reason is **approvals by standards organizations**. For example, PD and turbine flowmeters are approved by the American Water Works Association (AWWA) in the United States and the International Standards Organization (ISO) in Europe for use in custody transfer of water. The AWWA has approvals for both nutating disc and oscillating piston PD meters. While magnetic flowmeters are encroaching upon this territory, PD and turbine meters will continue to dominate the water custody transfer market.

The effect of approvals is shown by the historical example of AGA-9, which formulated criteria for the use of ultrasonic flowmeters for custody transfer of natural gas. After the publication of AGA-9 in 1998, the ultrasonic market for natural gas flow measurement received a major boost. Previous AGA publications laid out criteria for the use of DP and turbine meters and had a similar effect.

Users are also sticking with conventional meters because **suppliers are bringing out technologically advanced products**. Turbine suppliers are using materials such as ceramic to improve the life of ball bearings. Rosemount has introduced the 3051 S, a pressure transmitter with increased accuracy and stability. PD suppliers are using enhanced manufacturing techniques to achieve more precision for their PD meters. Communication protocols such as HART and Profibus are beginning to appear on turbine and PD meters. All these changes are resulting in improved, more reliable, and more versatile conventional meters for users to choose from.

## SWITCHING TECHNOLOGIES HAS A COST

While end users are not averse to changing technologies, they are not likely to do so unless they have a specific reason to make this change. One reason is having a problem with the flowmeter. Another is being bought out and having to go with the technology from a new company. Still another is budget requirements that dictate going to a less expensive meter. But changing technologies is not without cost. It usually means taking time to learn a new technology, finding a new supplier, and stocking a different set of backup parts. All these cost time and/or money.

Another reason why users continue to stay with conventional meters is that they are genuinely the best solution for certain types of flow applications. Each type of meter has its own set of applications in which it excels. This varies by meter type.

## SELECTING A FLOWMETER

Which flowmeter is best for a given application depends on the characteristics of the application. If the application includes very low flowrates or viscous flow, PD or thermal flowmeters are good choices to consider. For very high gas flowrates, a turbine or ultrasonic flowmeter is a likely choice. Every flowmeter has its own paradigm case applications, which include those in which it excels. Table 3.1 lists the applications in which the different types of conventional flowmeters excel.

**TABLE 3.1**
**Where Conventional Flowmeters Excel**

| Type of Flowmeter | Applications They Excel in | Disadvantages |
|---|---|---|
| Differential pressure | Clean liquids, steams, and gases without the need for highest accuracy | Pressure drop; orifice plates subject to wear |
| Positive displacement | Low flows and viscous flows | Moving parts subject to wear |
| Turbine | Steady, high-speed flows | Bearings subject to wear; limited ability to handle impurities |
| Open channel | Rivers, streams, partially filled large pipes | Medium accuracy |
| Variable area | Visual flow indication | Low accuracy; many without transmitters |

# 4 Differential Pressure Transmitters

## OVERVIEW

Differential pressure (DP) flowmeters rely on what is commonly called a "primary element" to complete the conditions for flow measurement. Primary elements represent a constriction in the flowstream that causes a drop in line pressure. A DP flowmeter uses the pressure differential thus created between points upstream and downstream of the constriction to compute the flowrate (Figure 4.1).

DP flowmeters have been around for more than 100 years, and they are one of the most studied and best-understood methods of measuring flow. Orifice plates are the most widely used type of primary element for measuring DP flow. Max Gehre received one of the first patents on orifice flowmeters in 1896. In 1909, the first commercial orifice flowmeter appeared, and it was used to measure steam flow. Soon after this, the oil and gas industry began using orifice plate flowmeters due to their low maintenance and a need for measurement standardization.

As orifice plate flowmeters became more widely used, several engineering organizations began studying their use. These included the American Gas Association (AGA), the American Petroleum Institute (API), the American Society of Mechanical Engineers (ASME), and the National Bureau of Standards (now known as the National Institute of Standards & Technology). In 1930, a joint AGA/ASME/NBS test program generated a coefficient-prediction equation. Tests performed at Ohio State University in 1935 in association with the NBS resulted in flow equations for DP flow that have been used by the AGA and ASME ever since. These equations are included in AGA-3, a report that has been updated a number of times since then.

Over the past several years, pressure transmitter suppliers have released a number of new products with advanced features. These features promise higher accuracy, greater reliability, enhanced self-diagnostics, and more advanced communication protocols. The promise of greater reliability is perhaps the strongest driving force behind the pressure transmitter market. Although some products may have a higher initial purchase price, end-users cite a number of reasons for shifting to higher-performing products. These include the need to conform to regulatory requirements, the need for reliability, a desire to standardize pressure products, and the need for custody transfer.

Some new transmitters also offer greater accuracy. Higher accuracy provides a reason to shift to higher performance for those users who are motivated by regulations, a desire to standardize, or the need to do custody transfer. End-users seem to be willing to pay for higher performance, although this varies with application and with features. Advanced communication protocols do not seem to be a major

DOI: 10.1201/9781003130024-4

**FIGURE 4.1** Orifice plate assembly in a pipeline and DP transmitter at an oil & gas site – Photo courtesy of Jesse Yoder.

drawing card for end-users, however, although some are contemplating an upgrade to HART.

The use of multivariable flowmeters has been growing substantially over the past several years. Several new companies have entered this market, including ABB, Schneider/Foxboro, and Yokogawa. Multivariable DP flowmeters measure more than one process variable, usually DP and/or pressure and temperature. Many of them incorporate a pressure and a temperature transmitter into a single device, making it unnecessary to order these products separately. Multivariable DP transmitters are mainly used to measure mass flow, and they are primarily used for steam and gas flow measurement.

Multivariable DP transmitters require a primary element to create the restriction in the flowstream that enables the DP flow measurement. They can be used with orifice plates, averaging Pitot tubes, Venturi tubes, and other primary elements. They are increasingly popular as end-users seek to reduce costs, and also to do more flow measurements of steam and gas. Multivariable flowmeters also provide more information about the process, and this is another attractive feature.

## DP AND OTHER PRESSURE TRANSMITTERS

This chapter is about DP transmitters that are used with primary elements to measure flow. DP transmitters are one type of pressure transmitter. Other types are gauge, absolute, and multivariable. Pressure transmitters assume the existence of pressure and transmit this pressure in the quantitative form to an instrument, device, or controller that is capable of acting on this value. Pressure transmitters include the following elements: pressure sensor; transducer, amplifier, or conditioning element; and output variable.

It is important to distinguish between pressure transducers and pressure transmitters. Pressure transducers are generally lower in cost and smaller than pressure transmitters and are typically not used in the process industries. They typically have loose wires at one end and do not perform at the same level as pressure transmitters.

DP flowmeters are composed of two elements: a pressure transmitter and a primary element. The primary element places a constriction in the flowstream thus creating a pressure change that makes DP flow measurement possible. The DP transmitter uses the upstream and downstream pressure values, together with Bernoulli's theorem, to compute flowrate. There are many types of primary elements including orifice plates, Venturi tubes, flow nozzles, Pitot tubes, wedge elements, and laminar flow elements. There are a large number of companies that only produce primary elements. They often leave it up to the customer to decide what kind of DP transmitter will be used with the primary element (Figure 4.2).

DP flowmeters have a huge advantage in the installed base, and end-users are in many cases staying with DP flow technology rather than switching to a different flow technology. End-users are likely to change technology only if their measurement requirements change in a way that calls for this, or if they have a problem with their existing technology. Many DP flow installations have been in place for many years, and they are operating without difficulty. If accuracy requirements change, however, this could prompt a switch to a new flow technology (Table 4.1).

**FIGURE 4.2** A DP transmitter mounted (top center of photo) on the framework above an orifice plate assembly in a pipeline.

**TABLE 4.1**
**Advantages and Disadvantages of DP Flow Transmitters and Primary Elements**

| Advantages | Disadvantages |
|---|---|
| Low cost | Low turndown/rangeability |
| Multivariable versions provide mass flow measurement for gas and steam | May have lower accuracy than alternatives |
| Different primary elements such as Venturis and flow nozzles create measurement flexibility | Difficult to provide diagnostics |
| A well-understood technology | Primary elements such as orifice plates are subject to wear and clogging |
| Have multiple approvals for custody transfer applications | The measuring point is within the flowstream |
| DP transmitters have become more stable and accurate | Less research is being done on new types of primary elements than on new-technology flowmeters |

## PRODUCT IMPROVEMENTS

Pressure transmitter suppliers have also introduced many product improvements, including greater accuracy and stability, which enable end-users to upgrade their technology without abandoning their DP technology. One important advance is enhanced accuracy in pressure transmitters, including DP transmitters. Other options include going to multivariable flowmeters or integrated flowmeters that contain both a DP flow transmitter and a primary element. In addition, suppliers have upgraded their Fieldbus and other communication protocol options to give end-users wide latitude in their communication options. All these options allow customers to stay with DP flow while opting for more technologically advanced products.

Bristol Babcock introduced the first multivariable pressure transmitters in 1992. Since that time, other companies have followed suit, including Emerson Rosemount, Yokogawa, and Honeywell. Multivariable transmitters measure more than one process variable in a single instrument. These transmitters typically measure DP, pressure, and temperature. In some cases, they produce a mass flow measurement, using these values. Multivariable pressure transmitters can also be used to measure level, as well as flow.

## MULTIVARIABLE TRANSMITTERS OFFER ENHANCED FLEXIBILITY

Multivariable transmitters can eliminate the need for a flow computer to perform the flow calculation. In some cases, the multivariable transmitter measures one or two pressure values and temperature, and then outputs these values to a flow computer that performs the flow calculation. In other cases, the computing power of the flow computer is brought onboard by the multivariable transmitter, which also performs the flow calculation. Emerson Rosemount has introduced a multivariable transmitter that includes an integrated primary element, resulting in a full-fledged multivariable flowmeter. The trend towards multivariable transmitters can be expected to continue in the DP transmitter and flowmeter markets. These products typically sell for less than it would cost to buy the transmitters or sensors separately, with an average selling price in the $2,000 range.

Besides offering a more economical way to measure flow, multivariable transmitters can measure mass flow as well as volumetric flow. This makes them useful for measuring steam and gas, in addition to liquids. While they do not achieve the same level of accuracy as Coriolis flowmeters, not all applications require Coriolis-level accuracy. There is a pronounced trend in the industry to bring the flow calculations onboard the flowmeter rather than having them done by a flow computer. This makes it easier to accurately calibrate the flowmeter. And if the multivariable flowmeter contains an integrated primary element, the entire flowmeter can be calibrated at once, with the primary element onboard.

## DP TRANSMITTER COMPANIES

### Emerson Automation Solutions

Emerson is a global technology and engineering corporation that manufactures and sells innovative products and solutions to industrial, commercial, and consumer markets. Emerson designs and manufactures electronic and electrical equipment, software, systems, and services for industrial, commercial, and consumer markets worldwide through two main businesses: The Automation Solutions business provides gas and liquid flowmeters and other products that automate and optimize production, processing, and distribution facilities for process, hybrid, and discrete manufacturers. The Commercial & Residential Solutions business offers products, systems, and software to enhance productivity, efficiency, and compliance (Figure 4.3).

#### History and Organization

The Emerson Electric Manufacturing Company was founded in St. Louis, Missouri, in 1890 by Charles and Alexander Meston, with financial backing by Wesley Emerson. Emerson Electric was initially a manufacturer of electric motors and fans, later launching sewing machines and washing machines. The company grew rapidly to nearly US$3 million by the end of World War I. The Great Depression caused a loss of nearly two-thirds of the company's sales, yet Emerson managed to survive. During World War II, much of Emerson's growth was a result of government contracts to build shell casings and airplane gun turrets.

Once the war ended, Emerson was able to begin commercial production again, but sales lagged and Emerson faced more challenging times. A different focus on creating foreign markets and research and development breathed new life into Emerson. From 1954 until 1973, the company grew from 2 plants to 84 plants, sales exploded from $56 to $800 million, and the employee count increased nearly 8-fold from 4,000 to 31,000.

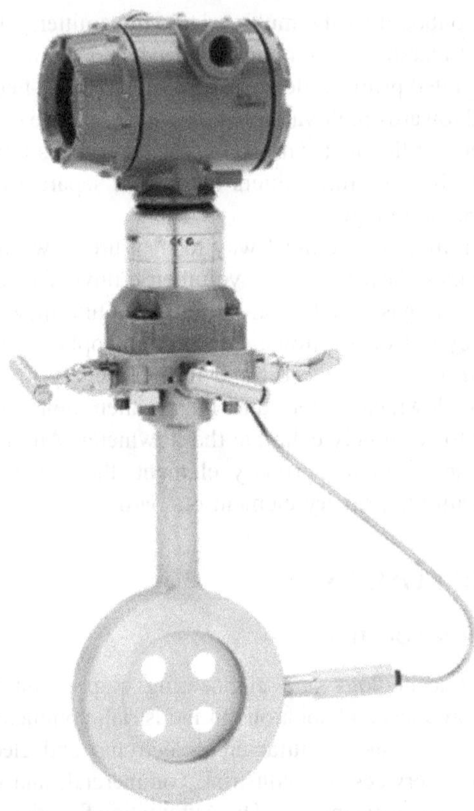

**FIGURE 4.3** Emerson's differential pressure Rosemount 3051SFC flowmeter comes with a patented, four-hole conditioning orifice plate – Photo courtesy of Emerson Automation Solutions.

Over the next 20 years, the company expanded through acquisitions (including Rosemount), joint ventures, new product development, and international growth. By 2000, Emerson's sales had reached US$15 billion, and the company was known as a global technologic leader that provided backup power for the telecommunications industry and built infrastructure to support Internet Protocol-based communications.

Today, the Automation Solutions division – home of Emerson's industrial process flow measurement products – is the next evolution of the business previously known as Emerson Process Management, which had evolved from the Fisher-Rosemount business. Emerson has retained its position as a recognized leader in process automation products and technology. Its flowmeter brands until recently included Micro Motion, Rosemount, and Roxar. On August 31, 2021, New York-based Turnspire Capital Partners announced that it had acquired the Daniel Measurement and Control business from Emerson. As a Turnspire Capital Partner Company, Daniel now offers turbine flowmeters, control valves, and primary elements, including the signature Daniel™ Senior™ Orifice Fittings. Emerson continues to offer Daniel's ultrasonic meters under the Rosemount brand.

All branded products have access to an extensive array of complementary accessory devices and networking schemes to choose from that – when combined – can create a synthesized system of functional usage and management control of a process. These integrated solutions include software, services, and other measurement and analytical instrumentation such as industrial valves, and process control software and systems.

Automation Solutions serves a range of industries: automotive; chemical; downstream hydrocarbons; food and beverage; industrial energy and onsite utilities; life sciences and medical; marine; mining, materials, and metals; oil and gas; packaging and filling; power generation; pulp and paper; and water and wastewater.

### DP Transmitter Products

Rosemount provides DP flowmeters in four basic series of design platforms: the 2051 and 3051 Series, the 4088 MultiVariable flow transmitter, and the most recent 9295 Process Flowmeter.

The 9295 Process Flowmeter is engineered as a complete integrated package and incorporates the company's proven conditioning orifice technology to eliminate straight run requirements. It can be specified with multiple transmitters for redundant measurements. Blockages in impulse lines can be removed without process shutdown. Optional temperature measurement reduces installation costs and allows for fully compensated flow measurement.

The 4088 Multivariable measures DP, static pressure, and process temperature, and comes equipped with Modbus as its communications protocol. The 2051 and 3051 Series are each available in wireless, compact, integral, and remote configurations with Annubar or orifice plate primary elements. The 3051 Series can be used for differential, gauge, or absolute pressure measurement, and is a scalable transmitter.

## Yokogawa Electric Company

Yokogawa Electric Corporation is a leading provider of industrial automation and test and measurement solutions, with 119 companies and operations in 61 countries. The company develops, manufactures, and markets information technology (IT) solutions, measuring and control equipment, semiconductors, and electronic components. Yokogawa's products include pressure transmitters, flowmeters, analyzers, data recorders, IT controllers, switching power supplies, and alternating current adaptors. In flow measurement, Yokogawa is the leading supplier of vortex flowmeters worldwide, and also provides magnetic, variable area, Coriolis, and DP flowmeters.

### History and Organization

Tamisuke Yokogawa, Doctor of Architectural Engineering, established an electric meter research institute in Tokyo in 1915, and Yokogawa became the first company to both produce and sell electric meters in Japan. Ichiro Yokogawa and Susumu Aoki established Yokogawa Corporation in December 1920. Recent executive appointments include the naming of Takashi Nishijima to serve as Chairman, and

Hitoshi Nara to serve as President and Chief Executive Officer. Yokogawa Corporation of America was established in 1957.

Yokogawa's industrial automation (IA) and control business – home of the company's flowmeters, pressure transmitters, and other field instruments – is by far the largest contributor to Yokogawa's bottom line, generating more than 90% of corporate revenue. This business is now optimizing solutions by focusing on three industry segments: Energy & Sustainability, Materials, and Life.

In addition to the industrial automation and control business, Yokogawa is developing the following two independent businesses: Measuring Instruments Business (high-precision technology for environment-related electrical power and optical communications) and New Businesses and Other (IoT hardware, software, and cloud solutions for service providers and an aircraft instruments business).

OpreX, the comprehensive brand for the IA and control business, consists of five categories: OpreX Transformation, OpreX Control, OpreX Measurement, OpreX Execution, and OpreX Lifecycle. OpreX Measurement field instruments include flow, multivariable, pressure, level, temperature, wireless, distributed temperature sensors, and handheld instruments. eX in OpreX stands for ***ex***cellence in the technologies and solutions that Yokogawa cultivates with its customers.

### DP Transmitter Products

Yokogawa offers three series of DP transmitters for liquid, gas, or steam flow as well as liquid level, density, and pressure. Yokogawa bases its pressure transmitter designs on its own "DPharp" (Differential Pressure High Accuracy Resonant Pressure) sensor technology.

The EJX-A series is Yokogawa's premium performance line of DPharp transmitters. The EJA-E series of transmitters is considered the workhorse of the company's DP transmitter product line, combining the ruggedness of the thoroughbred EJX-A series in a traditional-mount high gauge series. Both the EJA and EJX series operate using networked or optional wireless communication.

The P10 series can also be used to measure not only DP but also be equipped to provide flowrate, liquid level, density, and other process variables.

Yokogawa's multivariable transmitter, based on the higher-end EJX technology, combines a DP transmitter, a gauge pressure transmitter, a temperature transmitter, and a flow computer into one unit.

## HOW THEY WORK

A DP flowmeter consists of a DP transmitter integrated with a primary element and has the capability of calculating flowrate based on differences in pressure. DP flowmeters rely on a constriction placed in the flow line that creates reduced pressure in the line after the constriction. A primary element is used to create the constriction in the flowstream. A DP flowmeter also requires a means to detect the difference in upstream vs. downstream pressure in the flow line. While this can be done with a manometer, today's DP flowmeters use DP transmitters that sense the difference in pressure and then use this value to compute flowrate.

When most flowmeters are sold, the transmitter and sensor are sold together. This is true for ultrasonic, vortex, Coriolis, turbine, and other types of flowmeters. All these flowmeters operate based on a correlation between flowrate, or mass flow, and some physical phenomenon. For ultrasonic flowmeters, it's the difference in transit time of sound waves sent across the pipe. For turbine flowmeters, it's the speed of the rotor. DP flowmeters also correlate flow with a physical phenomenon; the difference in pressure upstream and downstream from a constriction in the flowstream.

Where DP flowmeters differ from other flowmeter types is that the transmitter is still often sold separately from the primary element that creates the constriction in the flowstream, sometimes from different suppliers. Because a DP flow transmitter cannot make a flow measurement without a primary element, customers who purchase a DP flow transmitter without the primary element are not actually buying a DP "flowmeter." They don't have a DP flowmeter until they connect the primary element to the DP flow transmitter. **Thus, a DP flowmeter is considered to be a DP flow transmitter that is connected to a primary element for the purpose of making a flow measurement.**

In the past, the pressure transmitter companies sold DP transmitters and users ordered their primary elements separately. Now, however, a number of companies in addition to Emerson Rosemount are selling DP transmitters already integrated with a primary element, such as an Annubar or an orifice plate. It is tempting to consider these the only true DP flowmeters; however, a better description is that these are DP flowmeters with an integrated primary element. If a customer assembles a DP flowmeter by connecting up a DP flow transmitter to an orifice plate or a Venturi tube from another source, the result is just as much a DP flowmeter as an integrated product.

## GROWTH FACTORS FOR THE DP FLOW TRANSMITTER MARKET

Although there are a number of negative factors at work in the DP flow transmitter market, the news is not all bad for suppliers. There are also a number of forces at work that promote growth within this market. These factors include the following:

- Energy efficiency and conservation
- Plant upgrades and retrofits
- Growth in developing markets
- The DP flow transmitter replacement market

### Energy Efficiency and Conservation

Energy conservation continues to be a major driving force in the manufacturing arena. One example of this is the growth of co-generation, where the steam output from power plants is used for other purposes such as heating. Conserving energy is part of a broader concern with reducing costs, which is driven by competition in the marketplace. Automation and enhancement of existing processes are two ways to enhance energy conservation. These automated processes are likely to incorporate flow measurement and control.

Managers of manufacturing plants are incorporating improved control of production processes to reduce waste and conserve energy. Better control of processes results in cost savings for customers. Many of these controls include flow measurement points, resulting in an increased need for flowmeters, including DP flowmeters. The need for energy efficiency and conservation will continue to increase, along with the increased cost of energy.

## Plant Upgrades and Retrofits

The process of retrofit and renovation is an ongoing process in manufacturing plants throughout the Americas, Europe, and Asia. Competition is a driving force behind the continuing need for companies to retrofit and renovate their plants. The prices of chemicals, petroleum products, paper products, automobiles, and other manufactured goods are always under pressure as consumers remain price conscious and as cost reductions by one company can force competitors to follow suit. In many cases, renovating a manufacturing plant can result in substantial cost savings. And when plants are renovated, companies are much more likely to install modern electronic equipment rather than outdated pneumatic equipment. Modern electronic systems are likely to incorporate more measuring points due to the need for greater efficiency and better control. As a result, plant renovations and upgrades are a real growth factor for the flowmeter market.

## Growth in Developing Markets

The construction of new plants occurring in the United States in many industries is less compared to construction of new plants in many other countries, especially in China. To meet the demands of its very large and growing population, China is investing heavily in plant construction. This includes chemical, pharmaceutical, power, water and wastewater treatment, food processing, and other process plants. These plants will require multiple measurements, including pressure, temperature, and flow, as well as control systems. While penetrating the Chinese market is not always easy, there is a great deal of growth potential there. Companies that have a broad instrumentation line, as well as the ability to deliver control systems, have a competitive advantage in China.

Plant construction is also occurring in many other regions of the world, including the Middle East, Africa, and Latin America. While some new plants are being built in Europe and the United States, the pace of new construction in these regions has slowed over the past several years due to the mature nature of these economies. Plant upgrades remain a source of DP flow transmitter sales in these regions, while larger orders can be expected to come from developing nations. However, with an improving economy in the United States, Europe, and Japan, comes hope for more DP flow transmitter business in these regions as well.

## The DP Flow Transmitter Replacement Market

The total size of the worldwide DP flow transmitter market is about 12% of the entire worldwide flowmeter market. And the DP transmitter market is large by any comparative measure. In fact, the DP flowmeter market is the third largest flowmeter market in terms of revenues – behind magnetic and Coriolis – when primary elements are included. But annual sales do not tell the whole story of the DP flowmeter market. The size of the installed base is a major reason why the pressure transmitter market will actually continue to hold its own within the instrumentation world.

DP flow transmitters have been around for more than 100 years, and this has resulted in a very large installed base for DP flow within the process industries. Because of the tendency to "replace like with like," many end-users can be counted on to continue to rely on pressure transmitters to make DP flow measurements. This means that they will continue to order new DP flow transmitters to replace DP flowmeters even where alternative technologies are available. In some cases, they will also order new primary elements when those are needed to maintain a DP flowmeter measurement.

## Higher Performance in DP Flow Transmitters

Over the past several years, pressure transmitter suppliers have released a number of new products with advanced features. These features promise higher accuracy, greater reliability, enhanced self-diagnostics, and more advanced communication protocols. The promise of greater reliability is perhaps the strongest driving force behind the pressure transmitter market. Although some products may have a higher initial purchase price, end-users cite a number of reasons for shifting to higher performing products. These include the need to conform to regulatory requirements, the need for reliability, a desire to standardize pressure products, and the need to do custody transfer.

Some new transmitters also offer greater accuracy. Higher accuracy provides a reason to shift to higher performance for those users who are motivated by regulations, a desire to standardize, or the need to do custody transfer. End-users seem to be willing to pay for higher performance, although this varies with application and with features. Advanced communication protocols do not seem to be a major drawing card for end-users, however, some of whom are contemplating an upgrade to HART.

# FACTORS LIMITING GROWTH FOR THE DP FLOW TRANSMITTER MARKET

In addition to macroeconomic factors, there are many factors specific to the pressure transmitter market that have an influence on growth in this market. This section looks at those factors. Factors limiting growth in the pressure transmitter market include the following:

- Pressure transmitters are more reliable, reducing the need for replacements.
- Other flow technologies are replacing DP transmitters.
- Multivariable transmitters include several transmitters in a single package.
- Market consolidation has led to product rationalization, reducing the availability of some products.

## Pressure Transmitters Are More Reliable, Reducing the Need for Replacements

Pressure transmitter suppliers have made major strides in reliability, including stability, in the past few years. One of the key features of Emerson Rosemount's 3051 Series pressure transmitter is greater stability, along with higher accuracy. Honeywell is now selling pressure transmitters with a lifetime warranty. Other suppliers have also increased the reliability of their pressure transmitters. Although greater reliability is very positive for end-users, it also means that they will have to buy fewer pressure transmitters to replace those that fail. This will have a negative impact on total pressure transmitter sales over time.

The long-term effects of greater reliability in pressure transmitters make the transmitter market somewhat like the market for automobile tires. When tires were made to last only 20,000 or 30,000 miles, car owners might have to buy two or three sets of tires during the lifetime of one automobile. Now some tires last 50,000 miles or more, reducing the number of sets to buy to one or two. Even though longer lasting tires often cost more, the fact is that most people have to buy fewer tires today than they did 5 or 10 years ago. Likewise, as reliability increases for pressure transmitters, end-users will have to replace them less and less often. This applies to the entire pressure transmitter market.

## Other Flow Technologies Are Replacing DP Transmitters

DP transmitters are a traditional technology method of measuring flow. Although this gives DP a major advantage in terms of the installed base, it also has resulted in the replacement of DP transmitters with primary elements by new-technology flowmeters. New-technology flowmeters include Coriolis, ultrasonic, magnetic, vortex, multivariable DP, and thermal. Of these technologies, Coriolis, ultrasonic, and magnetic flowmeters are having the greatest impact on the DP flowmeter market. End-users are selecting new-technology meters because of their higher accuracy and reliability. This trend will continue for the foreseeable future.

## Multivariable Transmitters Include Several Transmitters in a Single Package

Bristol Babcock introduced the first multivariable pressure transmitters in 1992. Since that time, other companies have followed suit, including Honeywell and

Emerson Rosemount. Multivariable transmitters are ones that measure more than one process variable in a single instrument. Typically, these transmitters measure pressure, DP, and temperature. In some cases, they use these values to produce a mass flow measurement.

Multivariable transmitters can be used to reduce the need to buy a separate flow computer to perform the flow calculation. In some cases, the multivariable transmitter measures one or two pressure values and temperature, and then outputs these values to a flow computer that performs the flow calculation. In other cases, the computing power of the flow computer is brought onboard by the multivariable transmitter, which also performs the flow calculation. Emerson Rosemount has also introduced a multivariable transmitter that includes an integrated primary element, resulting in a full-fledged multivariable flowmeter. The trend towards multivariable transmitters can be expected to continue in the DP transmitter and flowmeter markets. These products typically sell for less than it would cost to buy the transmitters separately.

### Market Consolidation Has Led to Product Rationalization, Reducing Availability of Some Products

Market consolidation has led some companies to reduce or eliminate some pressure transmitter product lines in an attempt to gain greater manufacturing and distribution efficiencies. This is especially true for companies such as ABB and Siemens that have grown through acquisition. Although the long-term effects of product rationalization may be positive, it may have a negative short-term effect because end-users who "replace like with like" may have to select a different type of pressure transmitter for replacement purposes. This can lead them to another manufacturer, or may simply lead them to order a different product from the same manufacturer. In the case of DP transmitters, it could lead them to explore alternative flow technologies. Thus, market consolidation may have at least a short-term negative effect on the pressure transmitter market. Of course, in some cases, it may only result in a shift from one supplier to another, rather than leading end-users to substitute a different technology than pressure altogether. The purchase of Bristol Babcock by Emerson may not have a negative effect because so far Bristol does not appear to be reducing its offerings as a result.

## FRONTIERS OF RESEARCH

The following are the frontiers of research for DP transmitters.

### Multivariable Flowmeters

Multivariable transmitters can be used to reduce the need to buy a separate flow computer to perform the flow calculation. In some cases, the multivariable transmitter measures one or two pressure values and temperature, and then outputs these values to a flow computer that performs the flow calculation.

In other cases, the computing power of the flow computer is brought onboard by the multivariable transmitter, which also performs the flow calculation. Emerson Rosemount has also introduced a multivariable transmitter that includes an integrated primary element, resulting in a full-fledged multivariable flowmeter. The trend towards multivariable transmitters can be expected to continue in the DP transmitter and flowmeter markets. These products typically sell for less than it would cost to buy the transmitters separately, with an average selling price in the $2,000 range.

Besides offering a more economical way to measure flow, multivariable transmitters can measure mass flow as well as volumetric flow. This makes them useful for measuring steam and gas, in addition to liquids. While they do not achieve the same level of accuracy as Coriolis flowmeters, not all applications require Coriolis-level accuracy. There is a pronounced trend in the industry to bring the flow calculations onboard the flowmeter rather than having them done by a flow computer. This makes it easier to accurately calibrate the flowmeter. And if the multivariable flowmeter contains an integrated primary element, the entire flowmeter can be calibrated at once, with the primary element onboard.

## Offering DP Transmitters with Integrated Flow Elements

The performance of DP flowmeters has been increased with the introduction of a primary element integrated with a DP transmitter. In the Emerson Rosemount Annubar Averaging Pitot Tube flowmeter product, a primary element with a DP flow transmitter is integrated to create a DP flowmeter. This reduces the need for impulse piping and valves, and also makes it possible to calibrate the device before shipping with the primary element already attached.

The popularity of these integrated flowmeters is likely to increase as end-users seek to cut costs and simplify the installation process. As these integrated flowmeters require a primary element, this is a frontier of research for DP transmitter companies.

## Technological Advances to Meet Advanced Plant Upgrade and More Stringent Environmental Requirements

More stringent environmental regulations and advanced plant upgrade requirements are making new demands on pressure transmitter suppliers. The need for more in-plant communication with other devices is resulting in the need for more communication protocols. The presence of wireless technology gives more flexibility in terms of where transmitters are located and a reduced need for wiring. Because a significant number of engineers are retiring, there is a need to build more intelligence into the transmitter itself. This includes advanced diagnostics and enhanced capabilities onboard the transmitter. Pressure transmitter suppliers are taking their cue from the needs of their customers, and building more capabilities into the transmitters to meet these changing needs.

## Higher Performance in DP Flow Transmitters

Over the past several years, pressure transmitter suppliers have released a number of new products with advanced features. Besides enhanced self-diagnostics and more advanced communication protocols, there is a major emphasis on higher accuracy, enhanced stability, and greater reliability. The promise of greater reliability is perhaps the strongest driving force behind the pressure transmitter market. Although some products may have a higher initial purchase price, end-users cite a number of reasons for shifting to higher performing products. These include the need for reliability, a desire to standardize pressure products, and the need to do custody transfer.

Some new transmitters also offer greater accuracy. Higher accuracy provides a reason to shift to higher performance for those users who are motivated by regulations, a desire to standardize, or the need to do custody transfer. End-users seem to be willing to pay for higher performance, although this varies with application and with features. Many of these features can be classified under the heading of improved product performance, which will continue to be a major frontier of research for pressure transmitter suppliers.

# 5 Primary Elements

## OVERVIEW

Differential pressure (DP) flowmeters are made up of two elements: a pressure transmitter and a primary element. The primary element places a constriction in the flowstream and creates a difference in pressure that makes measuring DP flow possible. The DP transmitter uses the upstream and downstream pressure values, together with Bernoulli's theorem, to compute flowrate. Many types of primary elements include orifice plates, flow nozzles, Venturi tubes, Pitot tubes, wedge elements, and laminar flow elements. There are many companies that only produce primary elements. In many cases, these companies let the customer decide what kind of DP transmitter will be used with the primary element.

Pressure transmitter suppliers have made many product improvements, including greater stability and accuracy, which allow end-users to improve their technology without abandoning their DP technology. One important advance is increased accuracy in pressure transmitters, including DP transmitters. Other options include multivariable flowmeters that contain a pressure and temperature sensor to compute mass flow. Another option is integrated flowmeters that contain both a DP flow transmitter and a primary element. In addition, suppliers have upgraded their Fieldbus and other communication protocols to give end-users many choices in their communication options. All these options allow customers to opt for more technologically advanced products while staying with DP flow.

Besides the options available in DP flow transmitters, some new options are available in primary elements. This includes more types of orifice plates, products that combine two primary element types into one, and cone meters. Expect more work to be done in primary elements over the next several years. Truly imaginative work in primary elements could go a long way towards increasing the popularity of DP flowmeters.

Other factors are also fueling the growth of the primary elements market:

**The rise of crude oil prices since 2021 and early 2022 has revved up oil and gas exploration and production activity worldwide.** Postpandemic demand for travel, going to events, and buying consumer products has greatly increased the demand for refined petroleum products. This increase in demand has created an imbalance between supply and demand, and increased the need for more oil and gas production. This surge in production has increased sales of DP flowmeters, which are among the most widely used flowmeters in the industry. This is good news for primary elements, which suppliers combine with DP transmitters to create DP flowmeters.

**Environmental standards and regulations for monitoring flare gas and stack gas emissions.** In response to continuous emission monitoring (CEM) requirements,

DOI: 10.1201/9781003130024-5

primary elements companies developed averaging Pitot tubes that use measurements at multiple locations to compute flow for the entire pipe, duct, or stack.

**Increasing popularity of multivariable and integrated DP flowmeters.** The multivariable DP transmitter market is growing as new suppliers of these more sophisticated devices enter the market. In addition, integrated DP flowmeters in which the transmitter and primary element are paired at the factory and sold as a single unit are also gaining in popularity. Instead of having to calibrate a primary element with a pressure transmitter that is purchased separately, end-users receive a single integrated unit that has already been calibrated.

**Technological improvements.** Product enhancements designed to solve specific application needs include extreme high-pressure wedge elements, cone-based meters, single and dual chamber housings, and hybrid combinations (e.g., flow nozzle with elements of an averaging Pitot tube).

**Strong installed base.** DP flowmeters have a very large installed base that almost guarantees the primary element – and DP flow transmitter – market will continue to hold its own within the instrumentation world.

## HISTORY OF PRIMARY ELEMENTS AND DP FLOW MEASUREMENT

The history of DP flow measurement goes back to at least the 17th century, though the measurement of flow using nozzles goes back to Roman times. At the beginning of the 17th century, Torricelli and Castelli arrived at the concepts that underlie DP measurement today: that flowrate equals velocity times pipe area and that the flow through an orifice varies with the square root of the head. In 1738, Bernoulli developed his famous equation for flowrate calculation.

The development of primary elements for use in measuring DP flow also began at about this time. Pitot presented a paper on the use of the "Pitot tube" in 1732. Venturi published his work on the Venturi principle for measuring flow in 1797. However, Venturi's work was not developed for commercial application until 1887 when Clemens Herschel used Venturi's work to develop the first commercial flowmeter based on it. In 1898, Herschel published his paper, *The Venturi Water Meter*.

Max Gehre received one of the first patents on orifice flowmeters in 1896. The first commercial orifice plate flowmeter appeared in 1909 and was used to measure steam flow. Shortly thereafter, the oil and gas industries began using orifice plate flowmeters due to ease of standardization and low maintenance. In 1912, Thomas Weymouth, of the United Natural Gas Company, did experimental work on the use of orifice flowmeters to measure natural gas. Weymouth used pressure taps located 1-inch upstream and 1-inchdownstream of a square-edged orifice. The Foxboro Company licensed Weymouth's work and used it as a basis for building orifice meters shortly after this time.

The increased use of orifice meters captured the attention of several engineering organizations. These included the American Gas Association (AGA), the American Petroleum Institute (API), and the American Society of Engineers (ASME). The National Bureau of Standards (NBS) also became involved in this research. In 1930, a joint AGA/ASME/NBS test program was able to generate a coefficient-prediction equation based on extensive tests. In 1935, tests performed at Ohio State University in conjunction with the National Bureau of Standards served as the basis for flow equations that have been used by the AGA and ASME since that time.

Work in the United States was combined with European work in the late 1950s and resulted in the issuance of ISO Standards R541 for orifices and nozzles and R781 for Venturis. Standard R541 was issued in 1967 and R781 was released in 1968. At about the same time, an ASME Fluid Meters Research Committee began a study to reevaluate the Ohio State data and to add new data on coefficients. The results were issued in an ASME Fluid Meters Report in 1971.

J. Stolz proposed a universal orifice equation in 1975. His idea was to combine the Ohio State data into a single dimensionless equation that could be used for corner, flange, and D- and D/2 taps. He presented his equation in a paper in 1978. This equation appears in the ISO Standard 5167, published in 1980, which combines the previously published R541 and R781 standards into a standard for DP flow. The ASME Fluid Meters research Committee adopted the ISO 5167 standard in 1981. In 1995, this standard was developed into the ASME MFC-3M standard for all orifice, Venturi, and nozzle flowmeters.

## THEORY OF DP MEASUREMENT

The theory behind DP flowmeters is that energy is conserved when flow passes across or through a constriction in the pipe. A more exact statement of this theory is known as Bernoulli's principle, which states that the sum of the fluid's static energy, kinetic energy, and potential energy is conserved across a constriction in the pipe. One form of Bernoulli's principle for incompressible fluids is as follows:

$$v^2/2g + z + P/rg = \text{Constant}$$

where $v$ is the velocity of the fluid, $g$ is the acceleration constant, $P$ is the pressure, $r$ is the density, and $z$ is the elevation head of the fluid.

The equation of continuity formulates a relation between fluid flowrate and velocity for fluids that are incompressible. It can be formulated as follows:

$$Q = A_1 \times v_1 = A_2 \times v_2$$

where $Q$ is the volumetric flowrate. The product of $A_1$, the amount of an incompressible fluid that crosses the area at point 1 over some set amount of time, and $v_1$, the velocity at point 1, equals the product of $A_2$, the amount of fluid that crosses the area at point 2 over the same set amount of time, and $v_2$, the velocity at point 2. (This is also sometimes written as $A_1v_1 = A_2v_2$.) Even though the cross-sectional area of the pipe at point 1 is larger than the area at point 2 because the fluid is moving faster at point 2 the same amount passes in the same time (Figure 5.1).

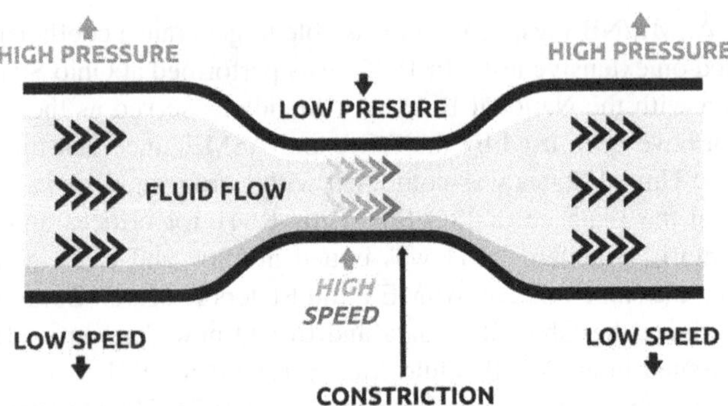


**FIGURE 5.1** Venturi effect.

Combining Bernoulli's principle with the equation of continuity yields the result that the DP generated by an orifice plate is proportional to the square of the flow through the orifice plate. This is the "square root" relationship that is fundamental to all orifice plate and other DP flow measurements.

## TYPES OF PRIMARY ELEMENTS

The types of primary elements are as follows:

- Orifice measuring points
- Pitot tubes
- Venturis
- Cone elements
- Flow nozzles
- Wedge elements
- Combination elements (e.g., nozzle/Pitot tubes)
- Others (e.g., Dall tubes, laminar flow elements)

### Orifice Measuring Points

Orifice plates are the most common type of primary element (Figure 5.2). An orifice plate is a flat, usually round piece of metal, often steel, with an opening or "orifice" in it. The orifice plate needs to be positioned at the correct location in the flow-stream for it to function as a primary element for the purpose of making a DP flow measurement. To be in position, it must be held in place. This is typically done by an orifice assembly, an orifice flange, or a holding element.

In addition to an orifice plate and assembly or flange, most orifice plate installations require the presence of a valve manifold, which serves to isolate the pressure transmitter from the process. DP flow transmitters use either a three-valve or a five-valve manifold.

As an orifice plate cannot serve as a functioning primary element unless it is held in the proper position, and since valve manifolds are required for most DP

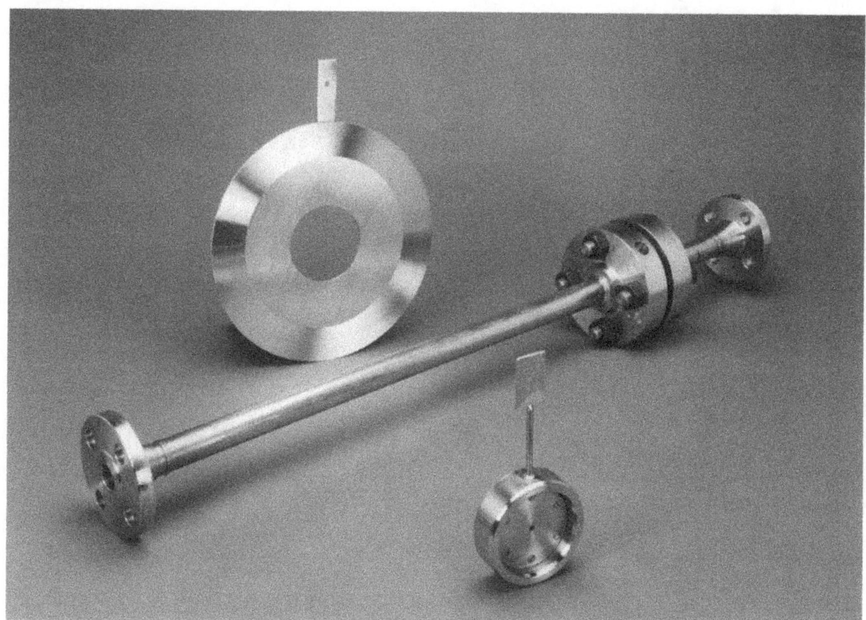

**FIGURE 5.2** Orifice plate, integral orifice flow element assembly, and restriction orifice plate – Photo courtesy of ABB.

flowmeter measurements, this book defines an orifice measuring point as having the following three components:

- An orifice plate
- An orifice assembly, flange, or holding element
- A valve manifold (Figure 5.3)

Orifice plates are classified according to the shape and position of the hole or opening they contain. The following are the main types of orifice plates:

- Concentric
- Conical
- Eccentric
- Integral
- Quadrant
- Segmental

## Pitot Tubes

The Pitot tube is named after Henri Pitot, who invented it in 1732. Henry Philibert Gaspard Darcy, another Frenchman, published a paper in 1858 that made improvements to Pitot's invention. The first patent for the use of a Pitot tube to measure velocity in pipes was given to Henry Fladd of St. Louis, Missouri, in 1889 (Figure 5.4).

**FIGURE 5.3** Orifice flange assemblies – Photo courtesy of ABB.

Pitot tubes are of two types:

- Single-port Pitot tubes
- Multiport-averaging Pitot tubes

A **single-port Pitot tube** includes an L-shaped tube that measures impact pressure. This tube is inserted into the flowstream, with the opening facing directly into the flow. Another tube measuring static pressure has an opening parallel to the direction of flow. Flowrate is proportional to the difference between impact pressure and static pressure.

A **multiport-averaging Pitot tube** has multiple ports to measure impact pressure and static pressure at different points. The DP transmitter computes flowrate by taking the average of the differences in pressure readings at different points.

Some companies such as Emerson Rosemount and Veris have introduced proprietary versions of the averaging Pitot tube. Emerson Rosemount's proprietary version is called the Annubar, and it was formerly sold by Dieterich Standard, now part of Emerson Automation Solutions. Veris' averaging Pitot tube is called the Verabar.

## Venturi Tubes

The Venturi tube was invented by an Italian physicist named Giovanni Battista Venturi in 1797 (Figure 5.5). In 1887, Clemens Herschel used Venturi's work to develop the first commercial flowmeter based on it. His version of the Venturi flowmeter became known as the Herschel Standard Venturi. Herschel published his

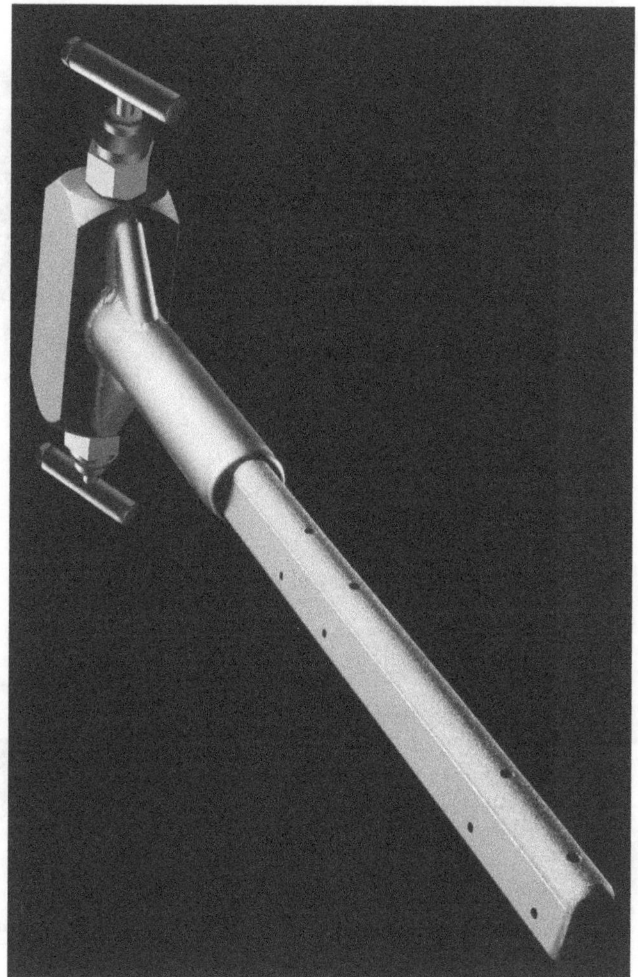

**FIGURE 5.4** Verabar Multiport Averaging Pitot Tube.

paper called "The Venturi Water Meter" in 1898. In 1970, a company called BIF introduced the Universal Venturi Tube™.

A Venturi tube is a flow tube that has a tapered inlet and a diverging exit. The DP transmitter measures pressure drop and uses this value to calculate flowrate.

## Cone Elements

Cone meters have been around since McCrometer, Inc., developed and patented the first successful model, its V-Cone®, in 1985 (Figure 5.6).

As illustrated below, cone meters consist of a specially tapered element positioned within the flowstream to create a restriction to the flow through the pipe.

**FIGURE 5.5** Venturi tubes – Photo courtesy of ABB.

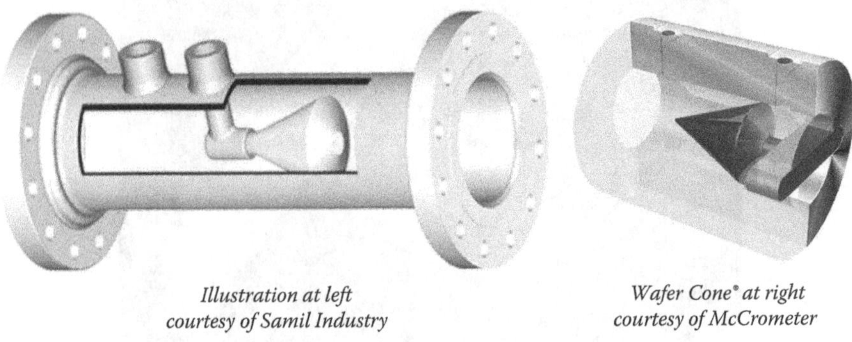

*Illustration at left courtesy of Samil Industry*

*Wafer Cone® at right courtesy of McCrometer*

**FIGURE 5.6** Cut-away views showing the position of the cone element.

Some cones are held in place by a tube that also connects them to the outside of the pipe. In the even more compact Wafer Cone® design, the element is fastened in place with an attached bar that spans the pipe diameter. The cone element creates a difference in speed and pressure as the flow is forced around it, then allowed to resume unobstructed flow beyond the cone. DP transmitters get measurements via a port in the pipe upstream from the cone (the high side) and also downstream (the low side) via another port either in the pipe wall or in the blunt end of the cone. Flowrate calculations are based on the same physical principles as other DP-type flowmeters, including Venturi meters. However, cone elements and meters are distinct from Venturi ones in design and are called just "cone" elements and meters. Perhaps, the confusion arose from the "V-" in the name, and people became familiar with the first and still-successful cone meters, McCrometer's proprietary V-Cone. But McCrometer says that "V-" actually refers to the shape.

Cone meters can be used to measure gas, steam, and liquids in a wide variety of industries. Extremely robust, accurate, economical, low maintenance, and useful in tight fits, they can be found from subsea to processing plants to Navy ships to satellites.

## Flow Nozzles

A flow nozzle is a flow tube with a smooth entry and a sharp exit (Figure 5.7). With, it might be said, elegant simplicity, the restriction caused by the reductions in diameter within the nozzle creates measurable changes in the flow through it, but the smooth, tapered shape causes less permanent pressure loss than an orifice plate.

The DP transmitter computes flowrate based on the difference between upstream pressure and downstream pressure.

Flow nozzles are mainly used for high-velocity, erosive, nonviscous flows. Flow nozzles are sometimes used as an alternative to orifice plates when erosion or cavitation would damage an orifice plate. They offer excellent long-term accuracy.

**FIGURE 5.7** A flow nozzle – Photo courtesy of ABB.

## WEDGE ELEMENTS

A wedge element is a flow tube that has a V-shaped flow restriction – the "wedge" – protruding into the flowstream from at least one side of the pipe (Figure 5.8). The wedge might be attached and contained inside the pipe or it might be created by heavy sections of metal welded into a notch cut in the pipe. In the latter case, there are often reinforcing bars across the notch (parallel to the length of the pipe). In any case, the wedge is solidly built in, making these elements extremely robust, with no moving parts, and no critical areas to wear or shift. They can handle any type of flow profile, whether laminar, transitional, or turbulent, and they are less prone to clogging or build up than some other element types. The wedge shape is usually symmetrical – presenting the same on the upstream and downstream sides – thus allowing for bi-directional flow. Or, in the case of highly erosive fluids, should there eventually be significant-enough wear on one side of the wedge, the flow tube can be turned around to present the unworn side to the flow thus prolonging the life of the meter. They are easy to install and easy to use.

There are variations of wedge elements designed to handle air and gases, steam, and all sorts of liquids – clean, dirty, high-solids, slurries, viscous, corrosive, or erosive.

**FIGURE 5.8** A cut-away view showing the position of the wedge in a COIN® Wedge Meter – Photo courtesy of Badger Meter.

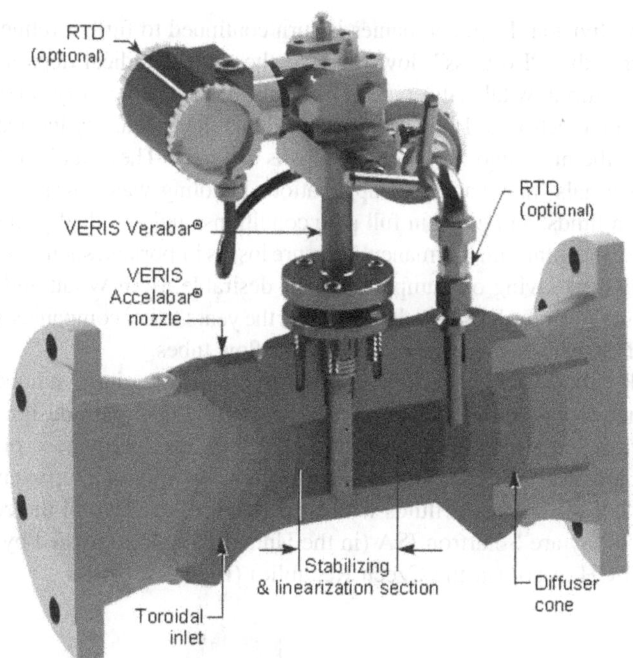


**FIGURE 5.9** A cross-sectional view of the Veris Accelabar® flowmeter that combines a specially designed flow nozzle and an averaging Pitot tube – Photo courtesy of Armstrong–Veris Flow Measurement.

## Combination Elements

An example of a flowmeter design that uses a combination of primary elements is the Veris Accelabar® flowmeter by Veris Flow Measurement (owned by Armstrong). In Figure 5.9, it can be seen how the flow enters the first element, a specially designed flow nozzle, and then encounters a second element, an averaging Pitot tube. The Veris accelabar nozzle's design provides a "settling distance" to accelerate, linearize, and stabilize the velocity profile of the flow. The Veris Verabar® averaging Pitot tube significantly increases the DP output and provides accurate measurement. The Accelabar's design makes possible a tremendous operating range (turndown) and eliminates the need for straight or upstream runs. It is able to cover an extensive range of applications.

## Other Primary Elements

Other primary elements include low-loss flow tubes, Dall tubes, and laminar flow elements.

**Low-loss flow tubes** are designed to produce a minimum amount of permanent pressure loss. Going back to the early 1960s, the first "Lo-Loss®" flow tubes were designed and marketed by the Penn Meter Company, bought by Badger Meter a few years later. In 2001, Wyatt Engineering purchased Badger Meter's differential-producing flow element division. Both companies in turn continued to further refine the "PMT" (Peter Meter Tube) "Lo-Loss" flow tube. At the time, a radical departure from the traditional Venturi flow tube, this type of flow tube has a continuously curved transition between the inlet section and the recovery cone rather than a long, cylindrical throat, and placement of the high- and low-pressure taps is different. They can be made in many sizes and materials for a variety of applications handling water, wastewater, sludge, slurries, clean fluids, and gases in full pipe conditions, and are ideally suited to applications where the minimum permanent pressure loss is important, such as in gravity-fed systems, or where saving on pumping costs is desirable. Both Wyatt and Badger still offer low-loss flow tube designs today and, over the years, other companies also began to make and offer their own variations of low-loss flow tubes.

**The Dall tube** was invented by an ABB hydraulics engineer named Horace E. Dall. It is an adaptation of the Venturi tube but with a flow path design that gives a shorter overall length, much higher DP, and very low permanent pressure loss. Drawbacks are that it is complex to manufacture, sensitive to turbulence, and not suitable for hot feed water or fluids with suspended solids. Two of the companies in this market today are Solartron ISA (in the United Kingdom, owned by AMETEK) and MATTECH, s.r.o. (in the Czech Republic) (Figure 5.10).

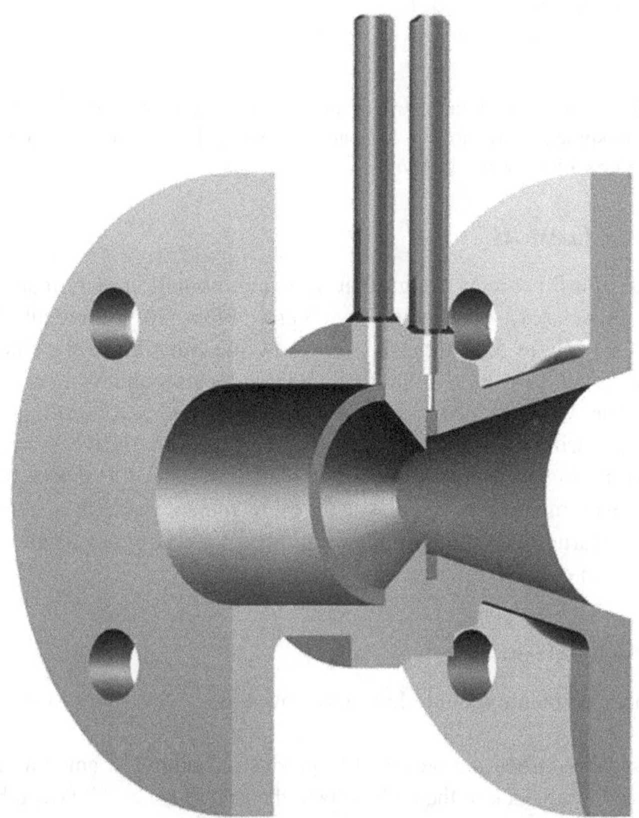

**FIGURE 5.10** Diagram of a Dall Tube – Diagram courtesy of MATTECH.

**Laminar flow elements** can be used for a wide variety of clean, non-condensing gas and air flows. (Note: In fluid dynamics, laminar refers to flowing in streamlines without turbulence.) Laminar flow elements can be made in sizes and materials to handle a wide range of flowrates and gases. They are used with mass flow controllers to create a pressure drop and a flow measurement. Their accuracy, stability, response time, and repeatability make them excellent for critical gas flow measurement applications as well as for calibration of various process instruments including some flowmeters, flow regulators, and thermal anemometers. They are also used to measure air flow to internal combustion engines. Other applications include leak detection, quantification, and testing. They are lower in cost than most other primary elements and somewhat difficult to quantify.

## PRIMARY ELEMENTS COMPANIES

### Daniel

Daniel, based in Houston, Texas, with a manufacturing facility in Chihuahua, Mexico, has been a global leader in fiscal flow and energy measurement to the oil and gas industry for more than 90 years. Until recently, Daniel was a brand within Emerson Electric's Automation Solutions business. On August 31, 2021, New York-based Turnspire Capital Partners announced that it had acquired the Daniel Measurement and Control business from Emerson. As a Turnspire Capital Partner Company, Daniel offers orifice fittings, liquid turbine flowmeters, and control valves.

#### History and Organization

Daniel was founded in 1946 by Paul Daniel, inventor of the Daniel Senior Orifice Fittings. Paul Daniel, born in 1894 in Houston and educated in a one-room schoolhouse, started his career in the oil industry in Southern California in 1915 at a refinery operated by Standard Oil Company.

Later, when out of work during the Great Depression, Daniel designed the now-famous orifice fitting to enable plates to be changed within the oil pipeline system without interrupting the flow of the oil or allowing significant leakage. A few years later he formed Daniel Orifice Fitting Company. By the time Daniel left active management in the mid-1960s, the company had more than 500 employees and the company's product line included piston-controlled check valves, orifice flanges, and Simplex plate holders. In 1966, the board of directors changed the company's name to Daniel Industries, Inc.

Emerson Electric Company, global technology and engineering company, acquired Daniel Industries for $460 million in 1999 to bolster Emerson's presence in the oil and gas industry. The Daniel brand within Emerson's Automation Solutions business segment aimed to deliver advanced insight and reliable measurement for the most challenging fiscal measurement applications. In addition to liquid turbine meters, the Daniel brand offered ultrasonic flowmeters and primary elements for both gas and liquid applications.

In February 2021, Lal Karsanbhai, who had served as executive president of the company's Automation Solutions business since 2018, became Emerson's CEO. He took over from David N. Farr, who was named CEO in 2000 and chairman of Emerson's board of directors in 2004.

On July 12, 2021, Emerson announced a definitive agreement to sell its Daniel Measurement and Control Business to Turnspire Capital Partners by the end of Emerson's fiscal year. The sale, announced as final on August 31, 2021, includes all of Daniel's brand rights, facilities, intellectual property, and personnel, but does not include Daniel's ultrasonic flowmeters, which remain with Emerson. Keith Barnard is the new CEO.

Turnspire invests in high-quality businesses that have reached "strategic, financial, or operational inflection points." The firm aims to make its companies best in class in their individual niches through operational improvements rather than financial leverage. Its stated mission is to "provide creative solutions to complex problems."

Turnspire's current portfolio includes Infinity Engineered Products, which engineers and manufactures Goodyear® rolling lobes, bellows, and sleeve air springs; MPI, North America's leading manufacturer of high-precision, fine-blanked metal components for automotive and industrial applications; UPG, an original design manufacturer and contract manufacturer of complex assemblies used in datacenter, automotive, energy, healthcare, and general industrial markets; Banker Steel Company, a full-service fabricator of structural steel components for commercial and infrastructure projects; and Crane Carrier Company, a leading manufacturer of purpose-built truck chassis and specialty engineered vehicles.

### Primary Elements Products

Daniel is a worldwide market leader in the production and sale of orifice plate equipment, including orifice plates and plate seals. These products are available in a wide variety of line sizes and materials, and are compatible with all Daniel fittings (i.e., Senior, Junior, Simplex, and Orifice Flange Unions). All Daniel orifice plates and seals are designed to ensure compliance with industry standards and to meet specific application requirements for custody transfer of oil and gas as well as allocation measurement. Daniel Venturi Tubes measure nonviscous fluids that contain hydrocarbons and other potentially corrosive elements, including LNG. Daniel Flow Nozzles are well-suited for applications with high-velocity, nonviscous, erosive flow.

## McCrometer

McCrometer, a Danaher company, supplies magnetic, V-Cone, propeller, Wafer Cone, and variable area flowmeters for liquid, steam, and gas measurement applications. The company prides itself on its stringent flowmeter calibration processes; each flowmeter is individually wet calibrated in one of the company's two world-class NIST traceable calibration facilities, one in California and the other in Nebraska, and delivered with a Certificate of Calibration.

### History and Organization

McCrometer was founded in 1955 by two brothers, Floyd and Lloyd McCall, who were unable to find the right flowmeter for their irrigation system and decided to design their own. Floyd had been asked by a group of water meter companies to develop a water flowmeter that was more accurate and reliable than the water flowmeters they were currently using. In response, McCall developed a propeller-type meter. The prototype of this meter is quite similar to the McPropeller water flowmeter that McCrometer still sells today.

McCrometer invented and patented the first successful cone meter, the V-Cone flowmeter, in 1985. Today, there are over 75,000 McCrometer V-Cone flowmeters installed worldwide.

Danaher bought McCrometer from Denver-based Ketema, Inc., in March 1996 for an undisclosed amount. In February 2000, McCrometer acquired Water Specialties of Porterville, California, the producer of the Water Specialties Propeller Meter and UltraMag line of magnetic meters. In March 2006, McCrometer further expanded its flowmeter product line with the acquisition of Marsh-McBirney's magnetic flowmeter line, the Multi-Mag, now known as the FPI Mag.

### Primary Elements Products

McCrometer's V-Cone flowmeter is used for liquid, steam, and gas in rugged conditions where accuracy, low maintenance, and cost are important. Due to built-in flow conditioning, the meter is inherently more accurate than traditional DP instruments such as orifice plates and Venturi tubes and is useful tight-fit and retrofit situations. For multiple clean water and wastewater treatment applications, the VM V-Cone is used. A prepackaged, built-in, three-way valve isolates the transmitter from the process fluid flow for easy maintenance.

The ExactSteam™ V-Cone System, factory configured for energy metering or mass flow, is a complete flowmeter for steam metering. It has a low-flow cut-off feature. The system acts as its own flow conditioner, fully conditioning and mixing the flow prior to measurement.

The low-cost, flangeless Wafer Cone combines exceptional flexibility with high performance. It is used in water & wastewater, chemical, food & beverage, plastics, pharmaceuticals, district HVAC, textile, power, and oil & gas production.

## GROWTH FACTORS FOR THE PRIMARY ELEMENTS MARKET

This section discusses the growth factors underlying the primary elements market. While there are some negative factors at work in the pressure transmitter market, there are also a number of forces at work that promote growth within this market. These factors include the following:

- The large installed base of DP flow transmitters
- Rapid growth in China, India, and other developing markets
- Technology improvements in primary elements
- Growth in the use of multivariable DP flowmeters
- Growth in the use of integrated DP flowmeters

## The Large Installed Base of DP Flow Transmitters

The total size of the worldwide DP flow transmitter market, including primary elements, is about one-fifth of the size of the worldwide flowmeter market. In fact, the DP flow transmitter market by itself, even without primary elements, remains one of the largest in terms of revenues. But annual sales do not tell the whole story of the DP flow transmitter market. The size of the installed base is a major reason why the DP flow transmitter market will continue to hold its own within the instrumentation world.

DP flow transmitters have been around for more than 100 years, and this has resulted in a very large installed base for DP flow within the process industries. Because of the tendency to "replace like with like," many end-users can be counted on to continue to rely on DP flow transmitters to make flow measurements. This means that they will continue to order new DP flow transmitters to replace DP flow transmitters even when alternative technologies are available. They are likely to consider making a change in technology only if they are having a problem with their DP flow measurement, or if specific application requirements change and DP flowmeters can no longer do the job.

Retention of DP-based flowmeters and their continued selection for service can be based on a number of interrelated factors. The familiarity of the basic technology means measurements are well understood by end-users and that this understanding is more easily transferred to other existing personnel or those who follow. Maintenance, diagnostic, and repair procedures have also become part of the mainstream routine. In addition, spare part inventories are easier to manage as relevant personnel have internalized product histories. And contractual obligations may also be based on the use of DP measurements in custody transfer and other allocation applications where there is a change in ownership of the fluid. Even given the growth of new-technology flowmeters, the newest data indicates that the use of DP transmitters is increasing, and, with it, the need for primary elements.

## Rapid Growth in China, India, and Other Developing Markets

The construction of new plants occurring in the United States in many industries is less compared to the construction of new plants in many other countries, especially in China and India. India is now the world's fastest-growing large economy, and probably will be for years to come. The International Monetary Fund has forecast GDP growth of 7.3% for India, with other forecasts also being within a very narrow range of that percentage. Many believe that China's maturing economy is unlikely to again grow at similar rates, rates that were common over a period of more than a decade.

To meet the demands of a large population with an increasing amount of discretionary income at its disposal, China is investing heavily in plant construction. This includes chemical, pharmaceutical, power, water and wastewater treatment, food processing, and other process plants. These are industries where primary elements have traditionally done well. These plants will require multiple measurements, including pressure, temperature, and flow, as well as control systems.

While penetrating the Chinese market is not always easy, there is a great deal of growth potential here. Companies that have a broad instrumentation line, as well as the ability to deliver control systems, will have a competitive advantage in China.

## Technology Improvements in Primary Elements

Primary elements suppliers continue to develop technological improvements, many of which are created as solutions to rather discrete application problems. For example, Primary Flow Signal stocks and supplies 35 different Venturi elements, many of which have been developed for a specific level of performance addressing customer applications that were not previously considered applicable for DP measurement.

Dosch Messapparate now includes in its averaging Pitot tubes in its primary elements portfolio as well as a new Cone meter and wedge elements that conform to ISO 5167-5 and ISO 5167-6 standards, respectively. Canalta Controls offers new Single- and Dual-Chamber Housings that ease the task of flow conditioner inspections, and make these inspections more cost-effective. Another recent addition to their portfolio is an internal valve isolation repair and/or installation kit. And this plug-and-play solution is suitable for use with all Canalta and Daniel chamber fittings, representing a sizable market base.

These recent innovations follow previous advances such as the Emerson Rosemount conditioning orifice plate that reduced the upstream requirements for DP flow measurement, and the Veris introduction of the Accelabar that combined elements of a flow nozzle with elements of an averaging Pitot tube. These developments together with others of more recent vintage have allowed DP measurement to more than hold its own in the worldwide market, keeping a trusted technology among the first-pick options for end-users across an array of industries.

## Growth in the Use of Multivariable DP Flowmeters

Multivariable DP flowmeters measure more than one process variable, usually DP and/or pressure and temperature. Many of them incorporate a pressure and a temperature transmitter into a single device, making it unnecessary to order these products separately. Multivariable DP transmitters are mainly used to measure mass flow, and thus they can be used for steam and gas flow measurement.

Multivariable DP transmitters require a primary element to create the restriction in the flowstream that enables the DP flow measurement. They can be used with orifice plates, averaging Pitot tubes, Venturi tubes, and other primary elements. They are increasingly popular as end-users seek to reduce costs, and also to do more flow measurements of steam and gas. Another attractive feature of these instruments is that they also provide more dynamic information about the process, enhancing management control and improving quality outcomes.

## Growth in the Use of Integrated DP Flowmeters

The performance of DP flowmeters has been increased with the introduction of a primary element integrated with a DP transmitter. In the Emerson Rosemount

Annubar Averaging Pitot Tube flowmeter, a primary element with a DP flow transmitter is integrated to create a DP flowmeter. This reduces the need for impulse piping and valves, and also makes it possible to calibrate the device before shipping with the primary element already attached.

The popularity of these integrated flowmeters is likely to increase as end-users seek to cut costs and simplify the installation process. As these integrated flowmeters require a primary element, this is a growth factor for primary elements.

## FACTORS LIMITING THE GROWTH OF THE PRIMARY ELEMENTS MARKET

This section discusses factors that may depress demand for primary elements. Other factors that serve to limit growth in the pressure transmitter market are as follows:

- Other flow technologies are replacing DP transmitters.
- In many cases, primary elements are commodity items.

### Other Flow Technologies Are Replacing DP Transmitters

DP transmitters are a traditional technology method of measuring flow. While this gives DP an advantage in terms of the installed base, it has not prevented the replacement of DP transmitters with primary elements by new-technology flowmeters. New-technology flowmeters include Coriolis, ultrasonic, magnetic, vortex, and thermal. Of these technologies, Coriolis, ultrasonic, and magnetic flowmeters are having the greatest impact on the DP flowmeter market. End-users have been selecting new-technology meters because of their perceived higher accuracy and reliability. This trend will continue for the foreseeable future.

More recently, purchasers are placing a higher priority on the suitability of a device within the new smart networking environment. Many new-tech flowmeter manufacturers have also introduced valuable levels of intelligence (e.g., self-diagnostics) within their devices themselves that were unavailable even a few years ago.

A suite of benefits that leading edge flowmeters offer include such features as remedy-based diagnostics, monitoring, and verification that not only satisfy regulatory, contractual, quality, safety, and fiscal concerns, but lower the total cost of ownership. The total cost of ownership is an even more critical buying consideration to users who may have previously only heavily weighted the one-time upfront cost of ownership. Increasingly, purchasers weigh the initial cost and long-term operational costs more equally.

### In Many Cases, Primary Elements Are Commodity Items

Some primary elements, such as those used for custody transfer of natural gas, are expensive and highly specialized items. But others, such as orifice plates by themselves, are low-cost commodity items. This means that prices are highly competitive for these products and there is a great deal of downward pressure on prices.

While this may not affect the extent to which orifice plates are used, it does make margin management more difficult for suppliers. As a result, rather than being dominated by a few large suppliers, the orifice plate market includes many low-cost providers who compete on price. This factor discourages some of the larger instrumentation companies from entering the primary elements market unless they have the advantage of patent protection or an extraordinary advantage in material or labor costs.

## APPLICATIONS

### Steam

Flow measurement of steam is dominated by two technologies: DP flowmeters and vortex meters. Between the two, DP flowmeters have the edge. One reason for this is that both flowmeter types can tolerate the high temperatures and high pressures of steam. DP transmitters can be mounted remotely from the process, and only the primary elements come in direct contact with the steam.

Steam applications cover a wide range of uses. Among the most popular steam flow measurements are boiler outlet, space heating, heating of process fluids, district heating, and steam injection.

Steam is often found in power plants to drive the generator turbines producing electricity. In this case, steam quality must be carefully monitored so that there is not too much water contained in it. When there is excess water content present, and if it is introduced to the turbines at high pressure and in immense volumes, the water causes an imbalance in the turbine and produces premature wear. Thus, flowmeters used in this type of electrical power generation must also be able to accurately measure flow even though the content of the fluid is subject to variability.

Regulatory requirements are a real driver for steam measurement accuracy in the nuclear power industry. In the United States, for example, the Nuclear Regulatory Commission oversees almost every aspect of a nuclear-based electrical generating facility. Oversight includes the areas of water inflows and outflows to the fission process itself, and this latter subject includes the supervision of steam usage. If the steam flows cannot be accurately measured and reliably reported as being within operational guidelines, then the plant's operation can be considered compromised and allowable electricity production either reduced or even curtailed entirely. If either of these actions is taken, there is an immediate effect on the operator's bottom line. Hence, nuclear facilities have a real need for accurate measurement of steam flow.

Another frequently found steam flow application is the re-use of process steam as a thermal source for environmental heating. The sense is that more firms are looking at introducing this application as the cost of heating and cooling facilities becomes more expensive.

In oil production, there is very active use of steam injection as a secondary extraction method of crude oil recovery. It is important for oil production companies to be able to maximize their recoverable reserves position. In many cases, only 25%–30% of the oil from a reservoir is taken out of the ground, primarily due to the lack of sufficient underground pressure to move the crude through the substrate to a point where it can be easily pumped to the surface. During the last decade when oil prices rose to the $90 and more per barrel range, exploration and production firms became extremely interested in using secondary recovery methods because the economics supported the use of such techniques. However, when oil prices dropped below $30 per barrel, those economies evaporated. During the first quarter of 2022, oil prices increased to above $100 per barrel. Considering that demand is outstripping supply, this price has staying power and is likely to sustain renewed interest in secondary recovery methods.

Another application is the use of steam in flaring. In this context, flaring is where steam is introduced within the exhaust stack on a tightly monitored basis as a way of reducing smoke. Accuracy is becoming more important here as stack temperatures are reduced due to either waste heat being recycled or simply as a result of steam inserted into the flaring process. If stack temperatures are reduced too greatly, the problem of proper draft becomes an issue.

### Measurement and Monitoring of Flare and Stack Gas Flow

Flare systems are used to burn off waste gases from refineries, process plants, and power plants. Flares can be a single pipe or a complex network of pipes. Flares are subject to strict environmental regulations. Flues typically are large pipes, stacks, ducts, or chimneys that dispose of gases created by a combustion process. Ultrasonic, averaging Pitot tubes, and thermal meters are used to measure flare and flue gas.

One of the biggest challenges of measuring flare gas is large turndown. Flare gas flow can range from low fuel gas purge during normal operation to large flow during emergency relief and/or total plant blowdown. Ultrasonic flowmeters can cope with a wide range of flow, and they also offer low-pressure drop, tolerate some condensed liquid, operate at high temperatures, and introduce no internal, insertion, or moving parts to block flare lines.

Gas flaring is now being more closely monitored worldwide because of its effects on the environment. Also, with natural gas becoming more important as a source of energy, natural gas that once was flared may now be captured and used as a source of energy. DP flowmeters using averaging Pitot tubes are one of the three main flow technologies used in gas flaring and stack gas measurement.

## FRONTIERS OF RESEARCH

The following are frontiers of research for primary elements.

### Technology Improvements in Primary Elements

Primary elements suppliers continue to develop technological improvements, many of which are created as solutions to rather discrete application problems. For example, Primary Flow Signal stocks and supplies 35 different Venturi elements, many of which have been developed for a specific level of performance addressing customer applications that were not previously considered applicable for DP measurement.

Dosch Messapparate now includes averaging Pitot tubes in its primary elements portfolio as well as a new Cone meter and wedge elements that conform to ISO 5167-5 and ISO 5167-6 standards, respectively. Canalta Controls offers new Single- and Dual-Chamber Housings that ease the task of flow conditioner inspections, and make these inspections more cost-effective. Another recent addition to their portfolio is an internal valve isolation repair and/or installation kit. And this plug-and-play solution is suitable for use with all Canalta and Daniel chamber fittings, representing a sizable market base.

## Developing More Types of Orifice Plates

Emerson Automation Solutions and other companies have brought out new types of orifice plates in the past 5 years, and this remains an area for further research. Types of orifice plates include concentric, eccentric, segmental, and quadrant edge orifice plates. In addition, conditioning orifice plates reduce the need for upstream piping. In some cases, these are offered with an integrated pressure transmitter, to form an integrated DP flowmeter.

Materials of construction are an important part of the effectiveness of orifice plates. Materials such as nickel and Monel make the plates more corrosion-resistant. The different designs of the orifice plates can enhance accuracy, and can also make the plate less susceptible to clogging by suspended particles in the flowstream. Even though orifice plates are probably the most studied type of primary element, new materials and new designs are a frontier of research for these very traditional forms of measurement.

## Combining Different Types of Primary Elements into One

While most types of primary elements fit into a single category, some companies have innovated by combining two types of primary elements into a new type of primary element. This is somewhat like a compound meter, which combines a positive displacement and a turbine meter into a single type of flowmeter. Compound meters are used in situations where the meter has to measure high flow sometimes and low flows at other times. For example, they are used in apartment buildings where water flow is high in the mornings and evenings, but very low in the middle of the night.

Armstrong–Veris offers the Accelabar, which is a combination of a flow nozzle and an averaging pitot tube. VorTek has introduced the VorCone, which combines elements of a vortex flowmeter, with elements of a cone meter. The initial tests and results on the VorCone are very positive. This combination of technologies makes it possible to predict the fluid density, volumetric flowrate, and mass flowrate without making use of any fluid density information from an external source. Among other benefits, this combination of technologies makes it possible for the VorCone to calculate the changing densities of gas mixtures.

While these are two examples, this is hardly the end of the story. Combining two technologies together can help to compensate for the weaknesses of a technology and enable it to address a broader range of applications. For example, vortex meters have a difficult time measuring low flows, because if the flow is low enough, the bluff body in the vortex meter will not generate the vortices required for a flow measurement. Pairing a vortex meter with a thermal meter could result in a meter that could handle medium to high flows, and also excel at low-flow measurement. While this is one theoretical example, there are no doubt other possibilities that will continue to be researched by innovative flowmeter companies.

# 6 Positive Displacement Flowmeters

## OVERVIEW

Positive displacement (PD) flowmeters are the workhorses of today's flowmeter world. They perform many important flow measurements that many people take for granted. For example, PD meters are widely used for billing applications for both water and gas. This includes residential, commercial, and industrial applications. PD flowmeters are widely used for water flow measurement at houses, apartments, and offices.

## PD FLOWMETERS ARE ONE OF THE EARLIEST TECHNOLOGY TYPES

PD flowmeters are among the founding members of the conventional flowmeter class. Their history goes back to 1815 when Samuel Clegg invented the first PD gas flowmeter. This was a water-sealed rotating drum meter. Clegg's son-in-law, John Malam, together with Samuel Crosley, invented an improved model in 1825. Problems remained, however, with high cost, freezing, and large size.

Thomas Glover invented the first "dry" gas diaphragm meter in 1843. Glover's meter contained two diaphragms and a sliding valve. In 1844, engineers Croll and Richards developed the first actual "dry" gas meter. The diaphragm meters used today are similar to these early meters, although major improvements have been made in the material of construction. Early meters had diaphragms made of sheepskin with steel metal enclosures. More recent meters have synthetic rubber-on-cloth diaphragms and are made of cast aluminum.

In 1824, Thomas Kennedy patented the reciprocating piston meter, sometimes called an oscillating piston meter. Nutating disc meters, used today as water meters, were invented in 1830 by James and Edward Dakwyne. The Dakwynes were granted a patent for a hydraulic pump using this same principle. In the 1850s, the nutating disc principle was incorporated into a meter developed by Bryan. These meters were improved, and the disc began to be made of hard rubber in the early 1900s. By combining hard rubber with brass, the life of the meter was greatly extended. This rubber and brass design was widely used until the late 1950s when the brass meter body and chamber were replaced by plastics and composites. Piston meters were first introduced in the early and mid-1800s. However, these meters were not very durable. The rotary piston meter was invented in the late 1800s, and it is still in use today.

Bopp and Reuther of Germany holds the earliest patent for the oval gear meter in 1932. This meter has since been popularized by Oval Corporation of Japan, which

DOI: 10.1201/9781003130024-6

introduced the oval gear meter in the early 1950s. Oval gear meters are used for liquid measurement.

The development of PD flowmeters was driven by the increasing need to accurately measure the amount of water and gas consumed in homes and in commercial and manufacturing establishments. PD meters predate the earliest new-technology flowmeter by more than 100 years. Magnetic flowmeters were the earliest new-technology flowmeter introduced, and this occurred in 1952. Because PD meters have been around so long, there has been more time for suppliers to develop them and build up an installed base. The very longevity of the PD meter helps to assure its continued use.

## UTILITY APPLICATIONS

PD meters are very effective at making low-cost mechanical measurements for utility purposes. They provide a very cost-effective solution for utility applications that require low-cost meters that last for many years in residential, commercial, and industrial utility applications. In these segments, the main competition for PD meters is from single-jet, multi-jet, compound, and Woltman turbine meters rather than from new-technology meters. One reason for this is that these turbine meters are relatively low in cost compared to new-technology meters. Also, industry approvals for new-technology meters such as magnetic and Coriolis have been slow to develop, although this is changing as they become more prevalent.

PD meters are often used as billing meters to measure the amount of gas used at houses, commercial buildings, and industrial plants. Industrial plants such as chemical, food processing, and pharmaceutical plants also use PD meters for billing purposes. These meters are different from the meters used to measure gas as part of the manufacturing process.

Automated meter reading systems are becoming more widely used to allow meters to be read from a remote location. While PD meters are losing out to Coriolis meters for some hydrocarbon measurements, and to magnetic flowmeters for some industrial liquid measurements, they still are the best solution for many applications.

Many of the PD meters used for gas utility measurements are diaphragm meters. However, these are being replaced by rotary meters for some applications, because rotary meters are smaller and lighter. In some cases, when customers take a diaphragm PD meter out of service, they replace it with a rotary PD meter. Rotary meters represent a newer technology, and they allow end-users to upgrade their measurement capability while staying within the PD class of meters. Rotary meters are also used for nonutility gas flow measurements in industrial environments.

## PD FLOWMETERS WILL BE AROUND FOR MANY YEARS TO COME

PD flowmeters are conventional flowmeters that will be around for many years to come. Even though they face stiff competition from new-technology meters in some segments, they still remain the best solution for certain applications.

**TABLE 6.1**
**Advantages and Disadvantages of PD Flowmeters**

| Advantages | Disadvantages |
|---|---|
| Ability to make low-cost measurements for utilities | Have moving parts |
| Best in smaller pipe sizes | Require periodic maintenance |
| Can easily handle very low flowrates | May have problems with impurities in the flowstream |
| Accommodate high viscosity fluids | Less research and development on PD meters than on new-technology meters |
| Especially effective for petroleum liquids | There is some reduction in throughput due to the flowmeter being in the flowstream |
| Long-lasting | |
| Industrial models are highly accurate | |

PD meters are very effective at making low-cost mechanical measurements for utility purposes. These include residential, commercial, and industrial utility applications. Meters that are 1 inch or less in size are used for residential applications, while 1.5- and 2-inch PD meters are typically used for commercial and industrial utility measurements.

PD meters for gas applications also face some competition from turbine flowmeters. However, PD meters are mainly used for smaller line sizes, and most PD meters have line sizes somewhere between 1.5 and 10 inches. Turbine meters, by contrast, perform best with steady, high-volume flows. For this reason, turbine meters are more likely to be used for line sizes more than 10 inches. This is also the range where ultrasonic meters excel. Of course, some ultrasonic meters are used at lower line sizes also. While ultrasonic, turbine, and PD meters overlap in the 4- to 10-inch size range, PD meters still have an advantage in the lower sizes. Low flowrates are not a barrier to PD meters. For this reason, PD meters will continue to be used in the smaller line sizes to measure gas and liquid flow.

In the area of oil flow measurement, PD meters face a stiff challenge from new-technology meters. The main competition for PD meters for oil measurement comes from Coriolis meters. Because oil is a high-value product, end-users are more willing to pay the higher prices of Coriolis meters to measure its flow. PD meters are widely used to measure the flow of hydrocarbon products both upstream and downstream of refineries at custody transfer points (Table 6.1).

## PD FLOWMETER COMPANIES

### Dresser Utility Solutions

Dresser Utility Solutions, formerly Dresser Natural Gas Solutions (NGS), is a leading provider of metering, electronics, instrumentation, flow control, distribution repair products, and over-pressure protection devices to utility and industrial customers. The company has a directly supported sales presence in more than 100 countries. The Dresser Utility Solutions family of companies currently includes Dresser Measurement, Dresser Pipeline Solutions, Dresser Utility Solutions UK, Flow Safe, ANDCO, RCS, Texsteam, and Nibsco Automation.

Dresser Measurement provides rotary PD flowmeters, metering instrumentation, and test equipment for global natural gas distribution, transmission, and production. Its electronic instruments range from simple solid-state volume pulse output devices to advanced volume correction instruments, including its Micro Corrector family of products. Dresser Model 6 prover systems offer an industry-leading solution for field and shop measurement of meter accuracy.

Dresser Roots® placed the first rotary gas meters in the United Kingdom over 50 years ago and today Dresser Utility Solutions UK enjoys a large and mature installed base and offers rotary meters and pulse transmitters as well as complete solutions from the application, to design, to build and commissioning support.

## History and Organization

Dresser Utility Solutions traces its history to 1880 when Solomon Dresser started selling products to the petroleum industry near Bradford, Pennsylvania. In 1888, Dresser, who later became a member of the US Congress, obtained a pipe coupling patent on a device used to isolate oil below ground. Following Dresser's death in 1928, W.A. Harriman and Co., Inc. converted Dresser's firm into a public company. In 1944, Dresser purchased Roots Blowers and Meters, which was founded in 1854 when two brothers, Francis and Philander Roots, introduced the rotary PD blower and marketed it in Pennsylvania oil fields. In 1920, the brothers introduced the world's first rotary meter for gas measurement.

In 1998, Dresser Industries merged with Halliburton, Inc., but in 2001, the Dresser Equipment Group (DEG) separated from Halliburton to become Dresser, Inc., owned by an investor group consisting of First Reserve Corporation, Odyssey Investment Partners, LLC, and members of the existing DEG management team. At the time DEG consisted of Dresser Valve Division, Dresser DMD-Roots Division, Dresser Instrument Division, Dresser Wayne Division, and Dresser Waukesha Division.

In July 2009, Dresser, Inc., acquired the assets of iMeter B.V., a global supplier of natural gas metering equipment based in the Netherlands known for its innovative product design capabilities and engineering expertise. The company manufactured and marketed rotary and turbine gas meters, meter instrumentation, and meter calibration systems for the natural gas industry.

In 2011, GE acquired Dresser for $3 billion from First Reserve, a global private equity firm exclusively focused on energy to expand its energy and industrial offerings, particularly its gas engine portfolio. In 2017, when Baker Hughes merged with GE Oil & Gas, Dresser's NGS business became part of the oilfield services company. In 2018, Baker Hughes, a GE company, sold what became known as Dresser NGS for an undisclosed amount back to First Reserve. (Dresser Italia S.r.l., also known as Talamona, was not included in the sale.)

## PD Flowmeter Products

Dresser Measurement offers accuracy, dependability, and low maintenance in custody transfer measurement applications from the well to the burner with its rotary PD gas meter. The robust and reliable meters are manufactured in North America and used for commercial and industrial applications, as well as for lower volume high-pressure applications.

The meters feature an oil-free body and are designed to measure the volume of gases and gas mixtures from the well to the burner with a high degree of accuracy and over a wide flow range.

Dresser Measurement offers 13 sizes of 8C-56M rotary meters for commercial and industrial metering applications, including low flow. Its Series B line mount meters are designed for custody transfer in gas distribution, gas transmission, and gas production, as well as measurement for industrial in-plant applications.

## TechnipFMC

TechnipFMC, a leading technology provider to the traditional and new energy industries, was established in 2017 from the merger of FMC Technologies, Inc., and Technip. The company's offerings range from individual products and services to fully integrated solutions in subsea, onshore, offshore, and surface projects. It is organized into two business segments, Subsea and Surface Technologies. The company maintains corporate headquarters in London, England, and two operational headquarters, one in Houston, Texas, and the other in Paris, France, and has operations in 48 countries.

TechnipFMC, through its Measurement Solutions division, is a leading supplier of flow measurement and control products to the oil and gas industry. The company's products and engineering include a comprehensive range of blending, transfer, metering, production, and loading systems. These include improving shale and subsea infrastructures and operations to reduce cost, maintain uptime, and maximize oil and gas recovery. Its flowmeter portfolio features products from a proud heritage including Smith Meter® and F.A. Sening®. The company has a particular focus on subsea systems.

## History and Organization

### *Technip*

Technip was established in 1958 by IFP (Institute of French Petroleum) in Paris, France. During the 1990s, Technip acquired a number of companies that strengthened its technical expertise and expanded its product line. The 2001 merger with Coflexip, a subsea flowline company, established Technip as a global leader in the energy industry. In 2011, the firm acquired Global Industries to expand its offerings for complex projects from subsea to shore. In 2015, Technip and FMC formed an alliance, Forsys Subsea, which led to the 2017 merger of the two companies.

*FMC Technologies*

FMC Technologies traces its roots to 1884 when inventor John Bean developed a new type of spray pump for his orchard and started the Bean Spray Pump Company. Mergers in the late 1920s with makers of vegetable processing equipment and cannery machinery resulted in a name change to Food Machinery Corporation (FMC). In 1961, the company changed its name to FMC Corporation.

FMC has supplied measurement equipment to the oil and gas industry for more than 90 years. It traces its roots to the Erie Service Station Equipment Company of Erie, Pennsylvania, founded in 1922. Five years later, the company changed its name to Erie Meter System, Inc. The first manufacturer of electric service station pumps, the company also produced oil pumps and dispensers, air compressors and towers, and greasing equipment. In 1958, the A.O. Smith Corporation of Milwaukee, Wisconsin, bought Erie Meter Systems and merged its operations with its Los Angeles division, the Smith Meter Company, whose founder's work and patents would eventually become known as the "Smith Meter." The company consolidated manufacturing in Erie, Pennsylvania.

In 1976, Geosource bought the corporation and changed its name 2 years later to Flow Measurement and Control Division of Geosource. In 1984, Moorco International purchased the Flow Measurement and Control Division and renamed it Smith Meter, Inc., a Moorco Company. In 1995, FMC Technologies acquired Smith Meter, to enhance a portfolio of measurement equipment that already included Kongsberg Metering and F.A. Sening. FMC has continued to develop the Smith Meter turbine and PD flowmeters and manufactures all Smith Meter brand products in Erie. In 1999, FMC Technologies acquired the flow measurement assets of Perry Equipment Corp. (PECO) of Texas. The company's orifice gas measurement products complemented the Kongsberg ultrasonic gas meter and strengthened the Smith measurement systems business. FMC Technologies then reorganized the combined products and services as its Measurement Solutions division.

The origins of the six-path ultrasonic flowmeter design that later became the MPU 1200 go back to the early 1990s. This design was developed by Fluenta/CMR. Fluenta was created by CMR in 1985 as an industrial outlet for its inventions. Kongsberg Offshore was a part owner of Fluenta. In 1993, Fluenta/CMR delivered a prototype of the six-path design ultrasonic flowmeter to Statoil. In 1997, FMC acquired the ultrasonic technology from Fluenta, and delivered the first ultrasonic flowmeter to Statoil, then named the FMU 700. Soon after, FMC introduced the MPU 1200 version, which incorporated the second-generation electronics. A third generation of transmitter electronics for the meter, with enhanced signal processing, followed in 2002. This third-generation technology was developed to accommodate both liquid and gas ultrasonic meters and coincided with the introduction of FMC's first liquid ultrasonic flowmeters in 2002.

That same year, in 2002, FMC restructured the company into two separate, publicly traded companies: a machinery business (FMC Technologies), and a chemicals business (FMC Corporation). In 2007, FMC Technologies released its Smith Meter Ultra6 liquid ultrasonic meter for custody transfer of crude oil, enhancing its existing liquid ultrasonic offering for refined product measurement.

In October 2009, FMC Technologies acquired Multi Phase Meters AS (MPM), based in Stavanger, Norway. The high-performance MPM product line was combined with FMC's Increased Oil Recovery (IOR) portfolio to offer a broader range of solutions for oil and gas customers.

In 2012, FMC acquired Control Systems International, Schilling Robotics, and Pure Energy Services. Also in 2012, FMC formed FTO services, a joint venture with Edison Chouest to provide integrated vessel-based subsea services. In 2015, the firm introduced Forsys Subsea, a joint venture with Technip, to further advance their design and leadership expertise in subsea applications. On January 17, 2017, FMC Technologies merged with Technip to form TechnipFMC.

On February 16, 2021, TechnipFMC spun off Technip Energies to create two independent, publicly traded, industry-leading pure-play companies. Technip Energies, with 15,000 employees, is now a leading engineering & technology company for energy transition, with leadership positions in LNG, hydrogen, and ethylene as well as growing market positions in blue and green hydrogen, sustainable chemistry, and $CO_2$ management.

TechnipFMC, with the remaining 22,000 employees, is the largest diversified pure-play in the industry.

### PD Flowmeter Products

TechnipFMC manufactures rotary PD meters for midstream and downstream petroleum liquids. The company is especially well known for its Smith Meter PD meters, which were introduced in 1940 and have remained a mainstay in the oil and gas industry. The meters have experienced a constant series of improvements, including technology, materials, and the use of computer-aided design and simulation to improve reliability. The company replaced the analog counter with electronics and says the next step will be a smart PD meter with more diagnostics for preventive maintenance. Today there are more than half a million Smith PD meters in the field—downstream, midstream, and upstream—including some 50- to 60-year-old meters.

## HOW THEY WORK

PD flowmeters operate by separating the fluid to be measured into distinct compartments of known volume. As the liquid or gas passes through the flowmeter, the flowmeter repeatedly fills and empties these compartments. Flowrate is calculated based on how many times these compartments are filled and emptied. Different types of PD meters differ according to whether they are designed for liquid or gas, and according to the shape and size of the compartments involved.

While there are a number of different types of PD meters, the main types are as follows:

- Oval gear
- Rotary
- Gear
- Helical
- Nutating disc

- Piston
- Oscillating piston
- Diaphragm
- Spur gear

## Oval Gear

Oval gear meters are used to measure the flow of liquids. They contain two oval-shaped rotors with gears that rotate as the flow passes from the inlet to the outlet. The rotors are meshed together with gear teeth. Because the meshed gears seal the inlet flow from the outlet flow, a difference in pressure develops across the rotors that results in the rotation of the rotors. This rotation of the rotors is also facilitated by the flow of the liquid as it passes from the inlet to the outlet area. Newer designs use a servomotor to drive the rotation of the rotors. A servomotor contains a motor coupled to a sensor for position feedback (Figure 6.1).

## Rotary

Rotary flowmeters are used when high accuracy is desired. These meters are tolerant of small amounts of particles, but larger ones can get lodged between the rotor and the edge of the meter, which can affect meter accuracy. There are a number of different designs for the rotary meter. A rotating impeller (rotor) that contains two or more vanes divides the spaces between the vanes into discrete areas of known volumes. As the fluid enters the flowmeter it is captured in each of these areas of known volume, and then emptied through the outlet. The flowmeter counts how many times this is done and computes flowrate based on that value.

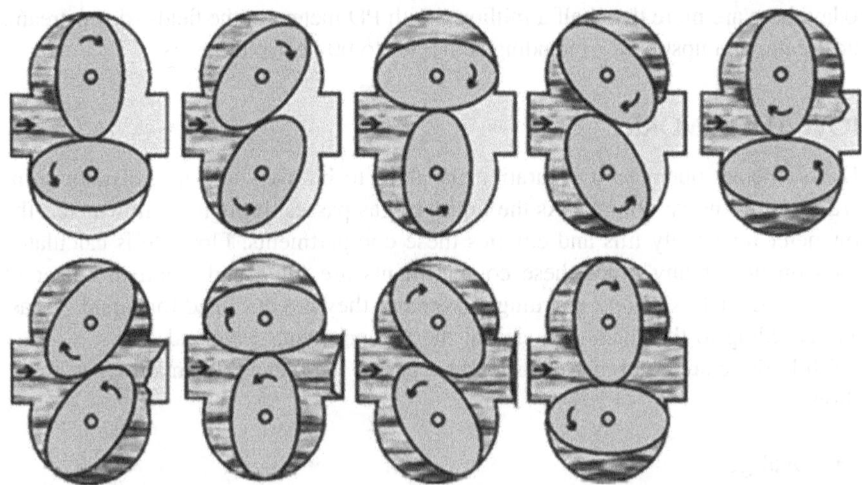

**FIGURE 6.1** Schematic of the function of an oval gear meter (gear teeth not shown) showing the start of inflow at the upper left to full outflow at the lower right – Courtesy of Bopp & Reuther.

Rotary flowmeters can achieve accuracy levels of ±0.1%. They are used to measure both liquids and gases and are used for a variety of fluids including crude oil, gasoline, and industrial liquids. Rotary flowmeters are available in sizes from 1 inch to 6 inches, with pressure ranges up to 150 psig. However, higher pressure ratings are available (Figure 6.2).

## Gear

Gear meters are used to measure the flow of liquids. They contain two rotors with gears that rotate as the flow passes from the inlet to the outlet. The rotors are meshed together with gear teeth. Because the meshed gears seal the inlet flow from the outlet flow, a difference in pressure develops across the rotors that results in the

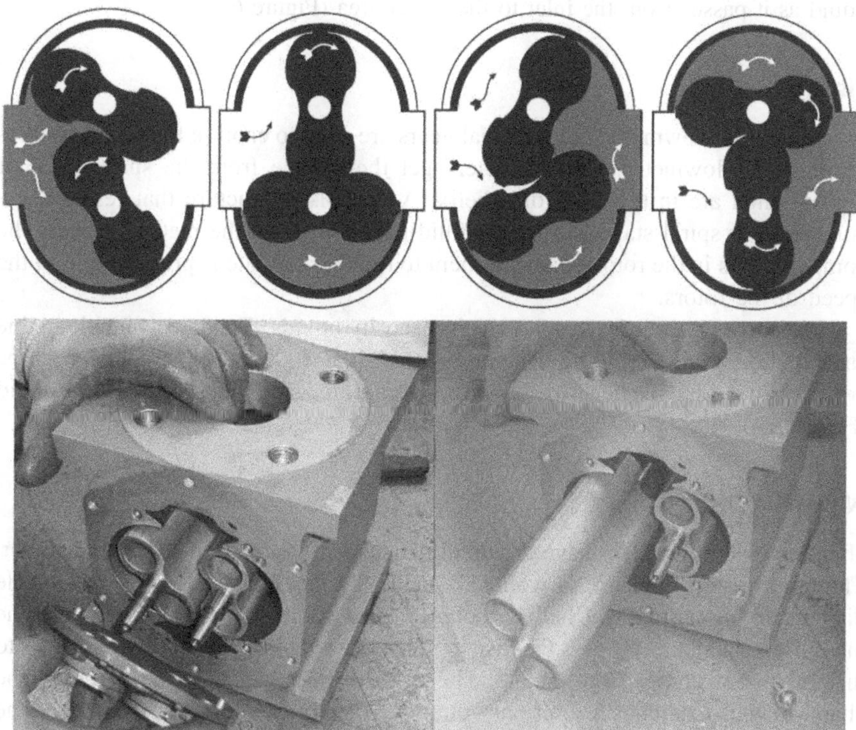

**FIGURE 6.2** Top: As shown in the cut-away illustration of a Dresser rotary meter, two contra-rotating impellers of two-lobe or "figure 8" design are encased within a rigid measuring chamber, with inlet and outlet connections on opposite sides. Precision machined timing gears keep the impellers in the correct relative position. Optimal operating clearances between the impellers, cylinder, and headplates provide a continuous, noncontacting seal. Bottom: Views of a partially disassembled meter in the Dresser Product Services Department show the impellers and measuring chamber inside the meter body – Courtesy of Dresser Measurement.

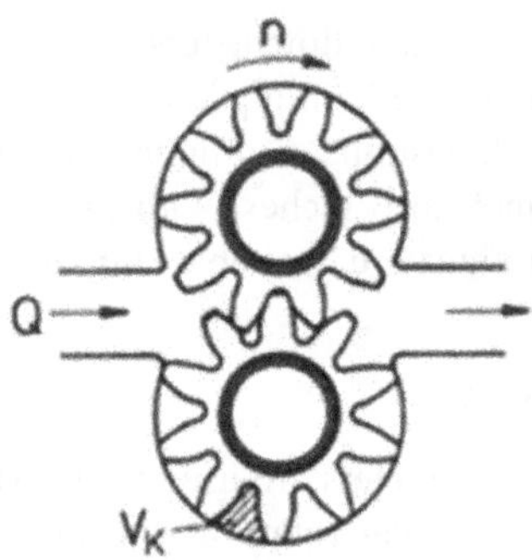


FIGURE 6.3 Schematic of gear positive displacement meter – Courtesy of Bopp & Reuther.

rotation of the rotors. This rotation of the rotors is also facilitated by the flow of the liquid as it passes from the inlet to the outlet area (Figure 6.3).

## Helical

In helical gear flowmeters, two helical gears are used to capture liquid as it passes through the flowmeter. Helical meters get their name from the shape of their gears, which are in the form of a helix. A helix is a structure that resembles a corkscrew or spiral staircase. As the fluid passes through the meter, it enters the compartments in the rotors, causing them to rotate. Flowrate is proportional to the speed of the rotors.

Helical meters are often used on high-viscosity liquids. They are available in line sizes of 1.5 inches, and turndown can be as high as 100:1. Accuracy levels can be better than 0.5%. Helical meters are tolerant of small particles in the fluid (Figure 6.4).

## Nutating Disc

Nutating disc flowmeters have a cylindrical-shaped base and measuring chamber. The term "nutate" means "to wobble." Inside the measuring chamber, a movable disc encircling a spherical ball is mounted on a spindle. As fluid flows into the measuring chamber, it causes the disc to nutate or wobble. This motion is similar to that of a coin dropped on a table. The space formed between the disc and the chamber wall maintains a constant volume as it moves within the chamber. The fluid in the measuring chamber is a fixed volume. As the disc moves, the motion of the spindle is translated through a magnetic assembly that can drive a register or a transmitter (Figure 6.5).

Nutating disc meters are widely used for residential service, although they are also used for commercial and industrial applications (residential meters are not included in this book). They are available in line sizes from 5/8 inch to 2 inches. Accuracy levels are in the range of ±2%, making them one of the least accurate PD meters. However, they also cost less than many other PD meters. Pressure is generally limited to 150 psig.

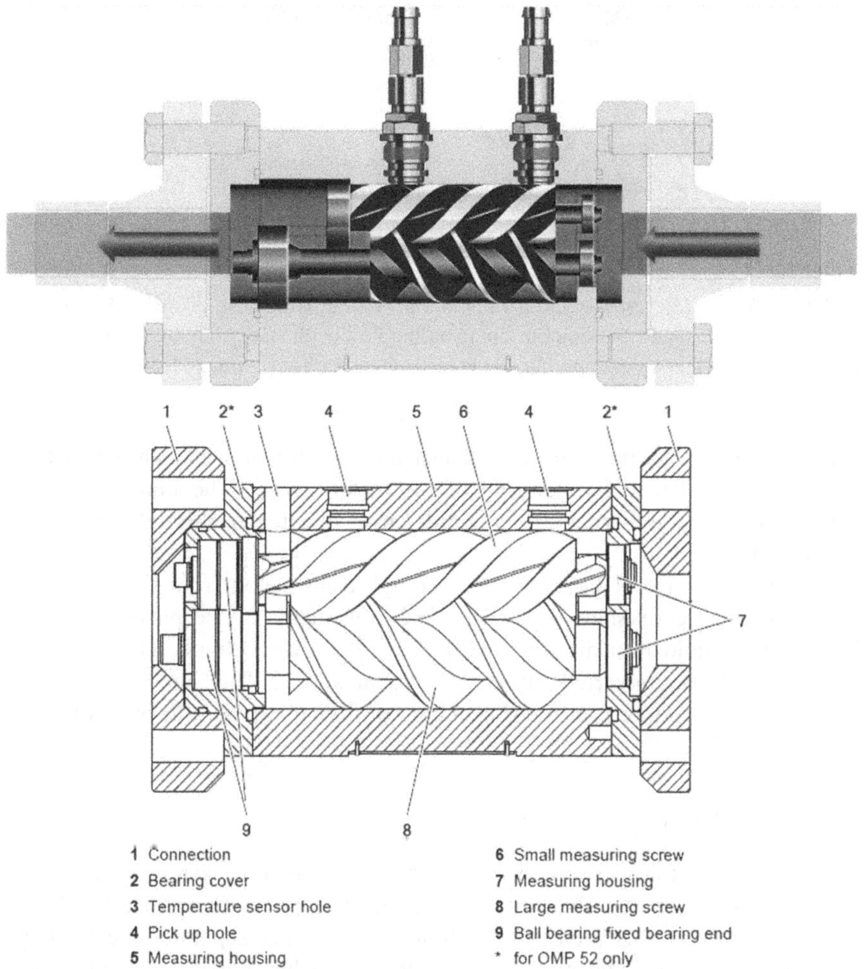


**FIGURE 6.4** Two slightly different cut-away views of KRAL OMP models illustrating same basic structure and precision workings of helical gear positive displacement flowmeters – Courtesy of KRAL.

## Piston

A piston PD meter uses four pistons to trap liquid as it passes through the meter. The pistons work somewhat like the pistons in a car engine and force a fixed amount of liquid through a cylinder. In the case of the piston flowmeter, the pistons are connected by a crankshaft that controls and synchronizes the movements of the piston. Liquid enters through the base of the meter and travels through the center cavity. As the crankshaft rotates and moves one of the pistons forward, a porting tube is connected to the center of the meter. As a result, liquid flows down the tube and displaces the adjoining piston. As the piston reaches the end of the cylinder, the porting tube is sealed off from the

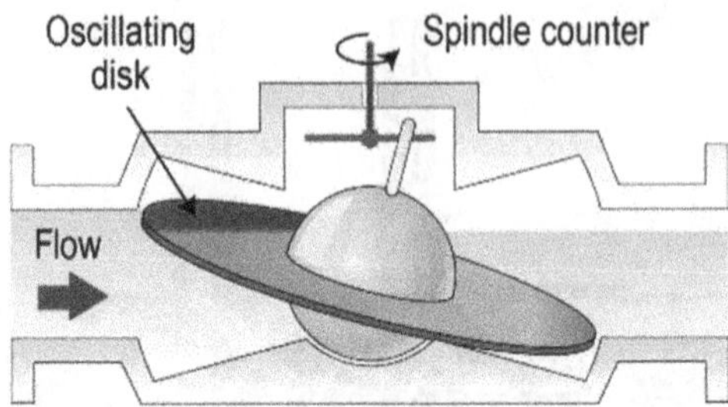


FIGURE 6.5 Middle cross-section illustration of a nutating disc positive displacement flowmeter. It is difficult to depict or envision the flow with static diagrams. This is meant to represent the right side emptying as the left side is refilling – Image from www.energy.gov/eere.

meter's center, exposing it to the exhaust port. Each piston has an exhaust port that is connected to the outer ring of the meter's base. The liquid from each piston collects in this ring. This liquid leaves the meter through the meter's exhaust port.

A known amount of liquid travels through each piston from the inlet to the outlet side. The crankshaft coordinates the movements of the pistons, each one dividing the liquid into discrete size pieces of known volume. The rotation of the crankshaft is converted into an electrical signal. This enables the piston flowmeter to display or transmit the amount of flow that passes through the meter (Figure 6.6).

## Oscillating Piston

An oscillating piston meter works somewhat like a nutating disc meter, although it doesn't nutate or "wobble" like a nutating disc meter. The cylindrical measuring chamber of this meter has a dividing plate that separates the inlet port from the outlet port. The piston is cylindrical and has holes in its supports so that liquid can flow freely on either side of the piston, and also on both sides of the plate that divides the inlet port from the outlet port. Rotation around a control roller guides the piston in the measuring chamber. The piston's motion is transmitted to a magnetic assembly that drives a register or a transmitter. The piston's motion is oscillatory. The flowmeter traps a discrete portion of fluid each time it rotates, so flowrate is proportional to the rotational velocity of the piston.

Oscillatory piston meters are made in line sizes from ½ inch to 3 inches. Their accuracy level is about ±0.5%. They are used as residential meters but are also used to measure the flow of clean viscous or corrosive liquids. Pressures should be 150 psig or less. The meter can easily tolerate the presence of small particles in the flowstream.

The oscillating piston meter is sometimes also known as the reciprocating piston meter. The reciprocating meter was patented in 1824 by Thomas Kennedy as a tool for indirect plumbing (Figure 6.7).

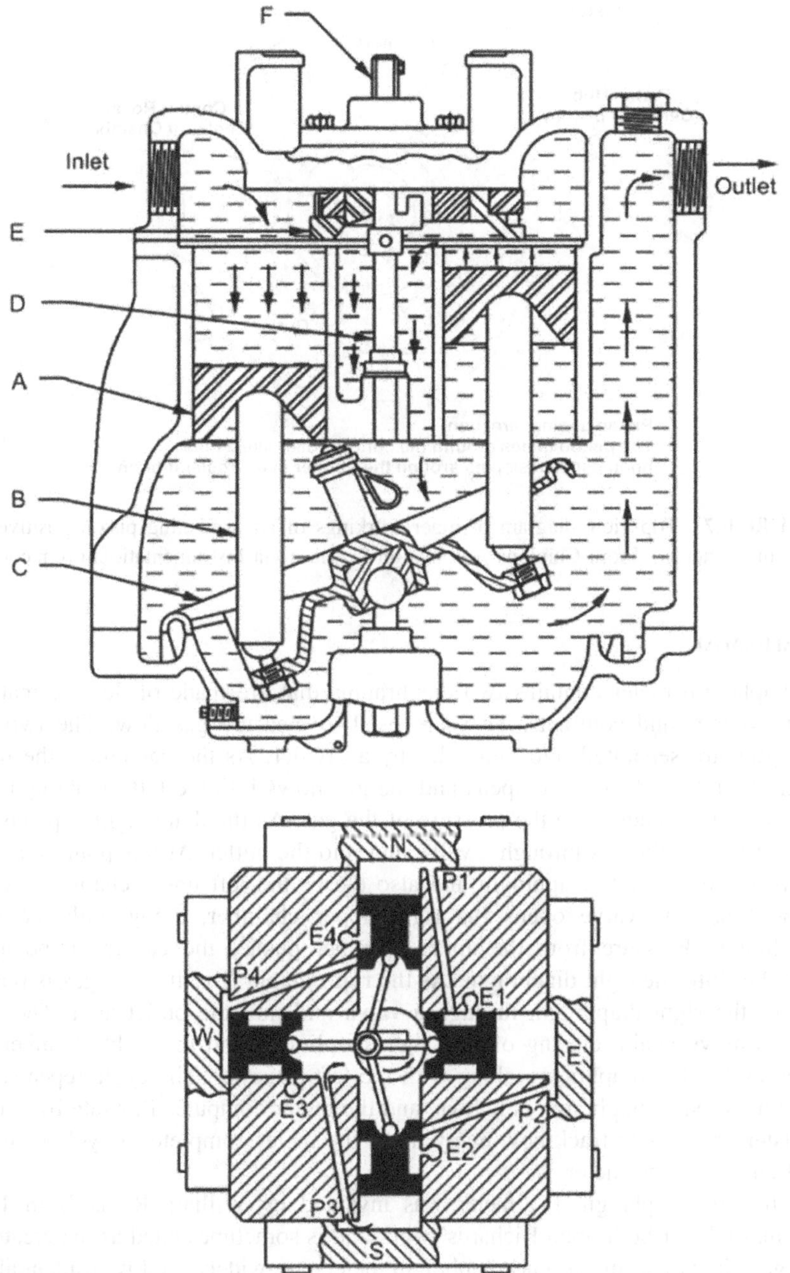


**FIGURE 6.6** Two variations of piston positive displacement flowmeters – From paktechpoint.com.

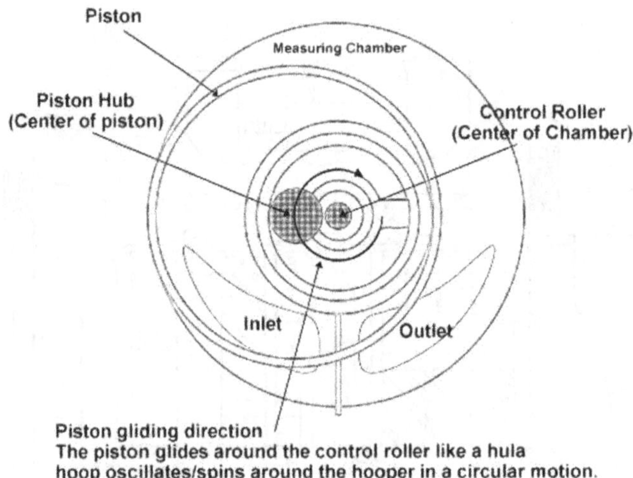


**FIGURE 6.7** Top view diagram of inner workings of an oscillating piston positive displacement meter – From Chipkins Automation Services via InstrumentationForum.com.

## Diaphragm

A diaphragm meter contains two diaphragms that are made of flexible material that expands and contracts. They are used to measure gas flow. The two diaphragms are separated into two sides by a divider. As the gas enters the meter from the inlet side a valve opens and the gas flows into the left diaphragm. The diaphragm expands from the pressure of the gas. As the diaphragm expands and fills, it pushes the gas through a valve and into the outlet. At this point, a sliding valve closes the left diaphragm and also closes the left upper chamber. At the same time, the valve opens the right upper chamber, along with the right diaphragm. Pressure from the upper chamber pushes the gas from the upper chamber into the right diaphragm. As the right diaphragm fills, the gas is pushed out of the right diaphragm through a valve and into the outlet area. The slide valves move again, closing off the right diaphragm and the right chamber and opening the left diaphragm, along with the left chamber. The cycle repeats. The volume of each diaphragm is known, and the meter computes flowrate by using a register or index to track how many times the meter completes a cycle of filling and emptying the meter.

The first diaphragm PD meter was invented by William Richards in 1843. Thomas Glover built upon Richards' work and is sometimes cited as the creator of the gas diaphragm meter. Gas diaphragm meters are widely used in residential and light commercial metering. Their accuracy level is in the range of ±1.0%, though some achieve higher accuracy than that. They are known for their ruggedness, and many have a life cycle of 10–20 years (Figure 6.8).

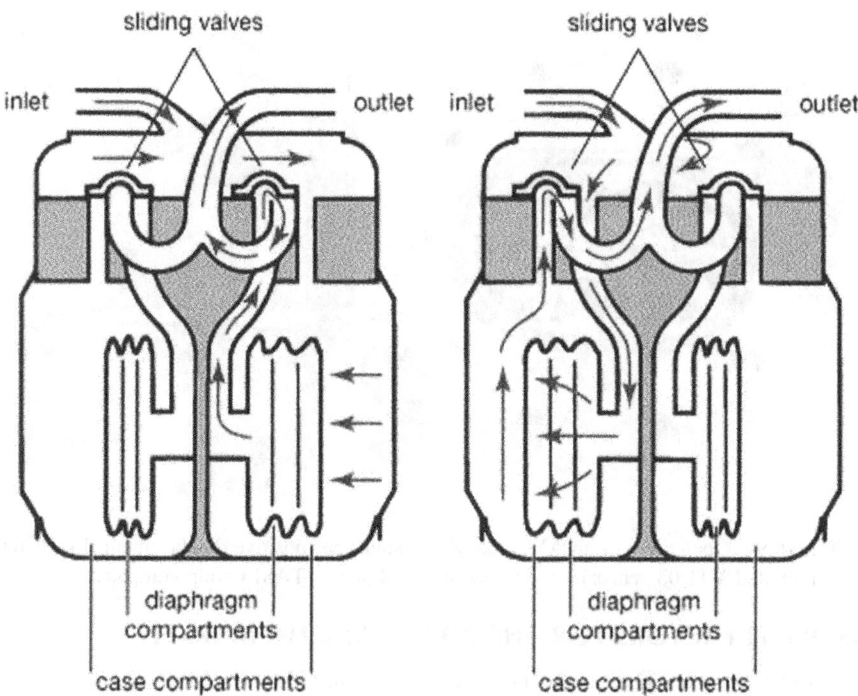


**FIGURE 6.8** Diagram of inner workings of one variation of diaphragm type positive displacement flowmeter – From camlockcouplings.blogspot.com.

## Spur Gear

Spur gear meters are another type of gear meter whose gears are arranged differently from oval gear and helical meters. Spur gear meters have two cylindrical gears that intermesh and directly face each other. Because of the way their gears are arranged, they produce high stress on the gear teeth and tend to produce high noise when operating. As a result, they tend to be used for lower speed applications, although they can be used at a variety of speeds (Figure 6.9).

## Other Types and Variations

So, we have seen that the basic principle of PD can be used in many different meter designs. In addition to those main types, and the variations illustrated, as covered above and by this book's segmentation, and there are sometimes other design variations based on each of those main operating principles, but we could not show them all.

There are also other less prevalent types of PD meter designs that fall, for this book, into the general category of "Other" and there are variations of those as well. A few of these other PD flowmeter designs are included in the sample drawing (Figure 6.10).

**FIGURE 6.9** Open view of an AW Lake ZHM spur gear positive displacement flowmeter (with mounted VTD03 sensor) – Courtesy of AW Lake (a TASI Group company).

## GROWTH FACTORS FOR THE PD FLOWMETER MARKET

Growth factors for PD flowmeters include the following:

- There are many PD flowmeter suppliers
- Large installed base
- New products, applications, and capabilities
- High accuracy a major factor
- Good for measuring low flowrates
- Excel with high viscosity liquids

### There Are Many PD Flowmeter Suppliers

Flow Research has identified nearly 100 suppliers of PD flowmeters worldwide. This compares to about 50 magnetic flowmeter suppliers and less than 25 suppliers of Coriolis meters. One reason for the large number of suppliers is that PD meters are used for residential as well as commercial and industrial applications. They are also able to measure both liquid and gas.

The major brands of PD meters have been around for many years. These include Dresser (once a part of GE and now within Dresser Utility Solutions), Smith Meter (now part of TechnipFMC), and American Meter (now part of Honeywell Elster). The former Roget Meter in Canada actually got its start due to a Dresser patent expiration. This created an additional company competing in the field of rotary gas meters. Today, given a large number of PD flowmeter suppliers, it is quite likely that more consolidation will occur.

When one company acquires another, it is usually to increase market share or profitability or to obtain new distribution channels. Because the PD flowmeter market is perceived by some to be in decline, PD flowmeter companies may be

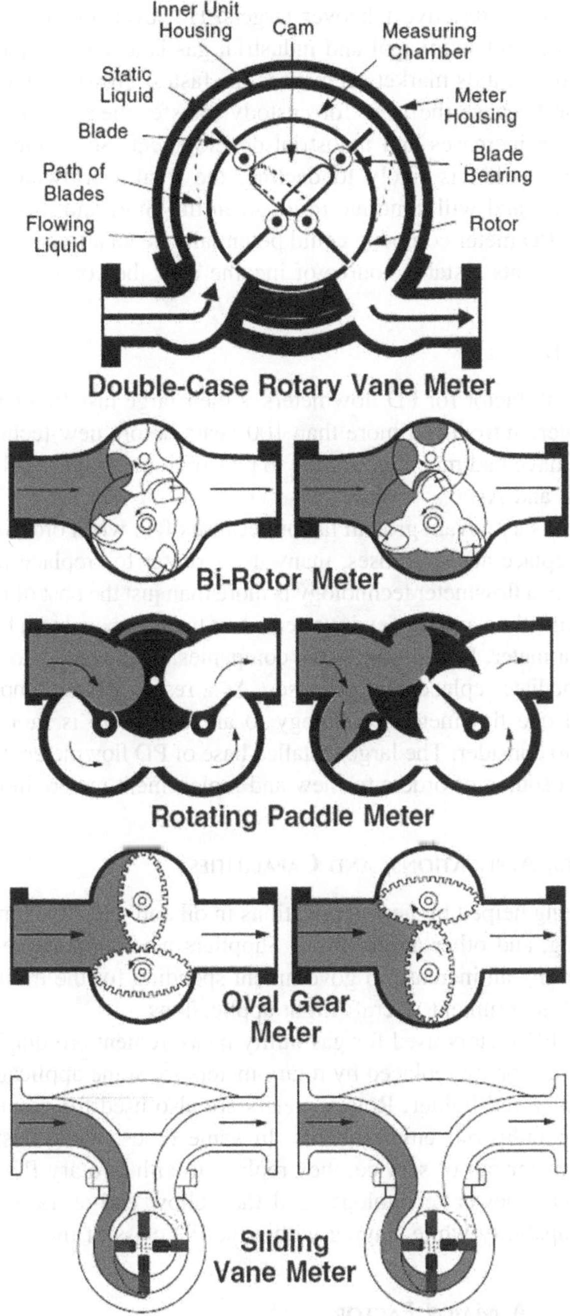


**FIGURE 6.10** Various positive displacement flowmeter designs – Courtesy of FMC Energy Systems.

perceived to be less attractive takeover targets. However, our research has found that the PD market for municipal and industrial gas is actually expanding slightly, and the petroleum liquids market segment is the fastest growing. This market has a very stable base because the need for custody transfer measurement is expanding due to population increases and industrial development. So, while the municipal/industrial water market is likely to decline, the total worldwide PD market is growing modestly and will continue to do so in the short and medium terms. For these reasons, a PD meter company could potentially be an attractive investment for a company that wants a stable source of income over the coming years.

## Large Installed Base

One major growth factor for PD flowmeters is their large installed base worldwide. Because they were introduced more than 100 years before new-technology meters, PD flowmeters have had much more time to penetrate the major markets in Europe, North America, and Asia.

Installed base is a relevant growth factor because often when ordering flowmeters, especially for replacement purposes, many users prefer to "replace like with like." The investment in a flowmeter technology is more than just the cost of the meter itself. It also includes the time and money invested when training people on how to properly install and use a meter. In addition, some companies stock spare parts or even spare meters for immediate replacement purposes. As a result, when companies consider switching from one flowmeter technology to another, there is more than just the purchase price to consider. The large installed base of PD flowmeters worldwide will continue to be a source of orders for new and replacement meters in the future.

## New Products, Applications, and Capabilities

PD sales are being helped by new applications in oil and gas, cryogenics, ultra-pure water processing, and other applications. Suppliers who support the aerospace industry are reporting an increase in government spending for the military as well as for civilian and government aircraft flight applications.

Many of the PD meters used for gas utility measurement are diaphragm meters. However, these are being replaced by rotary meters for some applications, as rotary meters are smaller and lighter. Rotary meters are also used for nonutility gas flow measurement in industrial environments. In some cases, when customers take a diaphragm PD meter out of service, they replace it with a rotary PD meter. Rotary meters represent a newer technology, and they allow end-users to upgrade their measurement capability while staying within the PD class of meters (Figure 6.11).

## High Accuracy: A Major Factor

Accuracy and reliability continue to be among the strongest factors driving demand in the flowmeter market. Some industrial PD flowmeters have accuracies in the 0.1% to 0.2% range. This high level of accuracy is equaled only by Coriolis and some multipath ultrasonic flowmeters. Some turbine meters also achieve very high

**FIGURE 6.11** PNG's natural gas PD meters in Surabaya, Indonesia.

accuracy levels. Popular residential and commercial utility PD meters are less accurate, but they are also much lower in cost.

PD meters are highly accurate because they actually separate the fluid to be measured into compartments and count the number of times this is done over time. PD flowmeter technology is the only flow measurement technology that directly measures the volume of the fluid passing through the flowmeter. There is no need for the inferential method that is used with meters that correlate flow with velocity or use the differential pressure (DP) method to measure flow. PD meters are widely used for custody transfer because they are both accurate and reliable. There are wide disparities in the degrees of both accuracy and reliability, however, depending on the manufacturer and the specific type of PD meter.

## Good for Measuring Low Flowrates

Unlike many other flowmeters, PD meters excel at measuring flow at low flowrates. Because PD meters actually divide the fluid up into smaller volumetric units and then count the units, low flow is not a problem for these meters. They can measure 1 gallon per hour or per day as easily as 1 gallon per minute or per second. For this reason, PD meters are widely used for residential and commercial utility measurements. Water volume in a home or a small business might be very high at certain times of day and slow to a trickle or be nonexistent at other times. Unlike other meters, including DP and vortex, PD meters have no problem with these low flowrates.

PD meters are more complementary than direct competitors to turbine meters. Turbine meters perform best with steady, higher flowrates. There is a group of turbine meters included in the single-jet and multi-jet categories that are used for residential applications, and another turbine type used here is the Woltman meter. However, in terms of industrial applications, turbine meters are more customarily used at the higher flowrates. For example, turbine flowmeters are widely used to measure the flow of natural gas for high-speed pipeline flows.

### Excel with High Viscosity Liquids

PD meters excel at measuring the flowrate of high viscosity liquids. For several viscous liquids that vortex meters and turbine meters have a problem with, PD meters do well. When PD meters are dividing the liquid flow by separating it into individual compartments, the viscosity of the fluid is of little consequence. There are many liquids in manufacturing that are viscous, such as honey and syrup. Magnetic flowmeters also have limitations in this area, such as in the measurement of animal fats and oils and vegetable oils. And, when compared to orifice-type meters used in the measurement of hydraulic fluids, PD meters perform well in that they require very little straight upstream piping because they are not sensitive to uneven flow distribution across the area of the pipe. PD meters are often the best solution for measuring the flow of all of these types of liquids.

## FACTORS LIMITING THE GROWTH OF PD FLOWMETERS

While there are many growth factors for PD flowmeters, there are also some limiting factors. Suppliers cite a saturated market and competition from new-technology flowmeters, especially Coriolis and ultrasonic. They also note that in US oil and gas markets, much investment is focused on production in the shale plays where PD meters are generally not used. A rise in the cost of materials is also affecting PD manufacturers' bottom line.

Factors limiting the growth of PD flowmeter market include the following:

- Competition from new-technology flowmeters
- Competition from conventional meters

### Competition from New-Technology Flowmeters

Competing technologies are gaining acceptance in some of the traditional PD markets. PD flowmeters face stiff competition from new-technology flowmeters, especially from Coriolis and ultrasonic meters, but also from magnetic meters. Coriolis meters are making inroads at both the microelectromechanical systems level and at the sub-one sizes. Suppliers maintain that new and alternative technologies, weigh batching systems, and foreign government regulations are negatively impacting PD sales. In fact, one manufacturer predicts that its own new magnetic meter solution for water will impact sales for its own PD meters.

PD meters face a stiff challenge from new-technology meters in the area of oil flow measurement. The main competition for PD meters for oil measurement comes from Coriolis meters. Because oil is a high-value product, end-users are more willing to pay the higher prices of Coriolis meters to measure its flow. PD meters are widely used to measure the flow of hydrocarbon products, especially downstream of refineries at custody transfer points and for fiscal and utility measurement.

Coriolis meters have also been developed for use in trucking and transportation measurement, LACT units, and also for in-plant measurement of hydrocarbons. While Coriolis meters are typically more expensive than PD meters, some users are willing to pay the difference in cost because of the high value of the product being measured. Magnetic flowmeters cannot meter hydrocarbons, so they do not compete with PD meters for oil measurement.

PD meters also face challenges from new-technology flowmeters in measurement for industrial applications. Both Coriolis and magnetic flowmeters are making inroads into measurement for these applications. Some users are seeking the high accuracy of Coriolis meters, whereas others are looking for meters that have no moving parts.

End-users of PD meters are familiar with the occasional need to keep the internals of their meter wet overnight or during extended off-times in order to help the next start-up. They are also familiar with maintenance routines such as disassembling, cleaning and lubricating gears, bearings and shafts, and installing and maintaining upstream filters to eliminate potential plugging. Or, where air pockets are common within the pumped fluid, installing air eliminators upstream of the meter to prevent damage and provide more accurate volumetric readings. These are all routines that new-technology flowmeters may not require.

## Competition from Conventional Meters

PD meters also face competition from turbine meters and compound meters. Single-jet and multi-jet turbine meters are widely used for billing applications for water, as are compound meters. Turbine meters are also used for custody transfer of natural gas. However, in certain respects, these markets are more complementary than competing. For instance, turbine meters perform well at steady, higher flowrates, whereas PD meters perform extremely well at low flowrates. This is why they are widely used to measure water for utility billing in homes and in commercial buildings. Flowrates in these buildings can be very low or nonexistent when no one is at home or in the buildings. So even though PD meters compete with turbine and compound meters, they still have their own areas in which to excel.

PD meters face some competition from turbine flowmeters for gas and other applications. One supplier, however, reported that some users are returning to PD meters after disappointment with turbine meters. PD meters are mainly used for smaller line sizes, and most PD meters have sizes somewhere between 1.5 and 10 inches. Turbine meters, by contrast, perform best with steady, high-volume flows and are more likely to be used for pipe sizes above 4 inches. This is also the range where ultrasonic meters excel. While ultrasonic, turbine, and PD meters overlap in the 4- to 10-inch size range, PD meters still have an advantage in the even smaller sizes. Low flowrates are not a barrier to PD meters. For this reason, PD meters remain a popular choice and are continuing to be used in smaller line sizes to measure gas flow.

## FRONTIERS OF RESEARCH

The following are some frontiers of research for PD flowmeters.

### From "The Meter with a Face" to Electronic Registers

Traditionally, PD meters have had mechanical register counters that resemble car odometers. These counters are incremented as the rotor inside the meter turns and indicates that a certain volume of fluid is passing through the meter. This counter gives PD a unique look that somewhat resembles a face. Of course, the purpose of the counters is not esthetic; it is to give an accurate reading of the fluid volume measured by the meter (Figure 6.12).

More recently, companies such as Liquid Controls and others are replacing these mechanical counters with electronic registers, or even flow computers. This opens up a whole new world of possibilities, including the addition of communication protocols, advanced totalization, higher accuracy, a visual display, easy configuration, and many other advantages. The electronic registers resemble the

**FIGURE 6.12** Digital register of a PD flowmeter.

programmable transmitters found on other types of flowmeters. This is a case where new technology is impinging on conventional meters, making them more accurate, reliable, and easier to use. Electronic registers are clearly a trend for the future in PD meters.

### Improved Manufacturing Methods Increase Precision

Some new developments focus on improved methods to increase the precision with which components are manufactured. For example, improved coordinate measuring machines make it possible to create more perfectly round pistons and other components. And as is the case with turbine flowmeters, improved bearing technology is making ball bearings used in PD meters more reliable and less prone to failure.

PD meters are benefiting from improvements in manufacturing technology that make it possible to increase the precision of the component parts of PD meters. The result is enhanced accuracy and reliability. Also using high-quality materials for construction results in longer life and higher accuracy in PD meters.

### Capitalize on Downstream Refined Fuel Distribution with Integrated Solutions

PD flowmeters are widely used for the downstream distribution of refined fuels. Much of this distribution involves PD meters integrated into systems with other equipment. For example, oval gear meters are incorporated into skids for use in the downstream custody transfer distribution of petroleum liquids. PD meters are also widely used for loading and offloading hydrocarbons in trucking terminals. Another use for PD meters is on the back of trucks for delivery of refined fuels. Here, they are incorporated into a system that typically includes pumps and valves.

Coriolis meters are making inroads into the PD market for some of these applications. However, PD meters still have a cost advantage over Coriolis meters, and they are highly accurate. One frontier of research for PD suppliers is to investigate how to make their meters more suitable for these OEM-type applications. This may involve designing meters that are more application-specific or even partnering with other companies that are involved in these markets. As regulations increase and auxiliary products become more sophisticated, developing application-specific solutions may be a way to minimize the encroachment of other flowmeters, especially new-technology meters such as Coriolis.

### Explore New PD Flowmeter Designs

This chapter explores at least eight different types of PD flowmeter designs. We have seen examples like the VorCone of flowmeters that combine two different flowmeter types into a single meter. PD meters are already combined with turbine meters to form compound meters that are designed for applications with widely varying flowrates. Perhaps by studying the advantages and disadvantages of different PD meters, it would be possible to come up with an improved design. One example of an improved design is that of rotary meters, which are replacing diaphragm meters for many applications. Rotary meters are more accurate, lighter and smaller, and more capable of handling higher volumes and pressures than diaphragm meters.

One route toward an improved meter would be to look for lighter and stronger materials for construction. Another is to have a design that minimizes or eliminates metal-on-metal contact. Still another is to maximize the durability of the bearings. It is also advisable to look beyond petroleum liquid applications to other industries, such as food & beverage. PD meters are used to measure the ingredients for making ice cream, and for measuring ingredients in the snack food industry. These are sanitary applications, unlike the petroleum liquid applications, and would require different approvals. Making flowmeters for specific applications is a tried-and-true approach to maintaining or gaining market share, and PD meters are no exception.

# 7 Turbine Flowmeters

## OVERVIEW

Turbine flowmeters are known for their robust strength and ease of use across a variety of gas and liquid applications. And, while the technology predates some other conventional types such as positive displacement (PD), their wider popularity is of more recent vintage. As one example, mid-20th-century engineers needed a reliable way to measure fuel on military planes during World War II and turned to turbine flowmeters. Their reliable performance in this difficult application helped bring them into wider industrial and commercial use.

Despite intense competition from ultrasonic, multiphase, and other new-technology flowmeters, turbine flowmeters have remained and will continue to be a viable and popular choice for a variety of applications. In particular, they excel at measuring clean, steady, medium- to high-speed flow of low-viscosity fluids. They also offer simplicity, effective turndown ratios, and the capability of customized solutions for various applications.

Turbine meters have a significant cost advantage over ultrasonic and Coriolis meters, especially in larger pipe sizes, although suppliers report increasing difficulty competing with ultrasonic and magnetic flowmeters in large line sizes. The price of turbine meters also can also compare favorably to DP flowmeters, especially in cases where one turbine meter can replace several DP meters. Users who are already familiar with turbine technology and do not want to spend the extra money required to invest in new technology are likely to stay with turbine meters.

## HISTORY OF TURBINE FLOWMETERS

While the Greeks and Romans had their own methods of flow measurement, turbine flowmeters are the first type of flowmeter of the modern era. Richard Woltman invented the first turbine meter in 1790. Woltman, an engineer, came up with the idea for the turbine meter while studying water loss in open canals. Woltman's name, along with his original design, persists today in the Woltman turbine meter. This meter is mainly used to measure water in bulk amounts.

Woltman's turbine meter did not rely upon a steady liquid fluid flow as did other measurement techniques available at the time. Over the decades, design development focused on measurement accuracy and linearity. It wasn't until the early 1940s that turbine meters began to be used for fuel measurement. Some of this development was due to the need to find a reliable way to measure fuel use on military planes used in World War II. Soon afterward, turbine meters began to be used in the petroleum industry to measure the flow of liquid hydrocarbons.

The use of turbine meters to measure gas flow began in 1953. Rockwell introduced a turbine meter to the gas industry in 1963, but it took about ten more

DOI: 10.1201/9781003130024-7

**TABLE 7.1**
**Advantages and Disadvantages of Turbine Flowmeters**

| Advantages | Disadvantages |
|---|---|
| Low to medium cost | Create pressure drop |
| Excel at measuring clean, steady, medium- to high-speed flows of low-viscosity fluids | Traditionally need manual lubrication |
| Good turndown/rangeability | Traditionally only uni-directional |
| Have AGA and API approvals for custody transfer applications | Bearings subject to wear |
| Reliable | Virtually no use for steam applications |
| Suppliers are bringing out new and upgraded turbine meters | |
| Excel at measuring the flow of high-speed natural gas | |

years for the gas industry to accept using turbine meters for measuring flow. In 1981, the American Gas Association published its Report #7, "Measurement of Fuel Gas by Turbine Meters." Since that time, turbine meters have been solidly entrenched in the gas industry as a measurement device, especially for custody transfer applications (Table 7.1).

Suppliers report that some customers are choosing electronic-based multipath ultrasonic meters over mechanical-based turbine meters as these are viewed as needing less maintenance and having nonintrusive designs. However, they also say that the higher costs of these meters and the high costs to calibrate them, combined with some uncertainty of the in-service accuracy and performance, are starting a trend back toward proven turbine meter performance – particularly dual rotor meters. Dual rotor designs offer improved accuracy and flow range, less pressure drop, and reduced flow swirl effects.

Turbine meter suppliers are making other technology improvements to make turbine meters more reliable. Many of these involve making the moving parts more reliable. By making the ball bearings out of more durable materials such as ceramic and sapphire, turbine suppliers have been able to add significantly to the life of the bearings. This is important since some customers select new-technology meters over turbine meters because turbine meters have moving parts that wear over time. Other recently introduced improvements include bi-directional flow, self-lubrication, significantly reduced pressure drop, and redundant meters.

## TURBINE FLOWMETER COMPANIES

### Honeywell Elster

Honeywell International is a Fortune 100 software-industrial company with over 1,300 sites in over 70 countries worldwide. Honeywell delivers industry-specific solutions that include aerospace and automotive products and services, energy, control technologies for buildings, homes, and industry, and performance materials globally. The company operates four business segments: Aerospace, Honeywell Building Technologies, Performance Materials and Technologies, and Safety and Productivity Solutions. Honeywell is listed on Nasdaq and the London Stock Exchange.

The Honeywell Performance Materials and Technologies division develops process technologies, software, and services for the Oil and Gas industries, as well as other advanced industrial devices. The division accounted for $9.4 billion of total 2020 Honeywell revenues. Within this division sits the Honeywell Process Solutions business, which provides industrial automation and control solutions – including pressure transmitters and flowmeters – for process and hybrid industries, including, refining, oil and gas, pulp and paper, mining, minerals and metals, bulk and batch chemicals, pharmaceuticals, and power generation. The Process Solutions group alone contributed $4.6 billion to Honeywell's gross revenue stream in 2021.

In December 2015, Honeywell International, Inc. finalized its purchase of Melrose Industries Plc's Elster unit for $5.1 billion. Today, Elster remains a world leader in the measurement of natural gas, electricity, and water with a product reach into more than 130 countries worldwide.

## History and Organization

Honeywell traces its roots back to 1885 when Albert Butz patented a furnace regulator and alarm. In April 1886, he formed the Butz Thermo-Electric Regulator Co. in Minneapolis. Shortly after that, a patent was granted for a device he invented called the "damper flapper."

The Consolidated Temperature Controlling Co. acquired Butz's patents and business, and renamed itself Electric Heat Regulator Co. In 1898, W.R. Sweatt purchased the company, later changed the name to Minneapolis Heat Regulator Company, expanded its product line, and patented the first electric motor approved by Underwriters Laboratories. Around the same time, an engineer named Mark Honeywell was perfecting the heat generator as part of his plumbing and heating business. In 1906, he formed the Honeywell Heating Specialty Co., Incorporated.

The company's name was officially changed to Honeywell, Inc., in 1963. In 1970, Honeywell merged its computer business with General Electric to form Honeywell Information Systems. By 1991, Honeywell was no longer in the computer business, but applied its knowledge to its traditional field of automation control, integrating sensors, and activators. In 1999, Honeywell was acquired by AlliedSignal, which retained the Honeywell brand name. The company's headquarters was moved to the AlliedSignal headquarters in Morristown, New Jersey.

From 2002 to 2017, the company acquired over 80 companies and divested itself of 60 businesses. In 2018, Honeywell streamlined its business by spinning off their Transportation Systems and Homes and Global Distribution businesses. Under new Chairman and CEO Darius Adamczyk (formerly COO), Honeywell is focused on growth, innovation, improved customer experience, and shareholder value.

## Honeywell Elster Group GmbH

The Germany-based Honeywell Elster Group GmbH is a worldwide market leader and specialist in flow measurement and control equipment and systems focused on the gas, water, and electricity industries in more than 130 countries. With one of the most extensive installed revenue measurement bases in the world and more than 200 million metering modules deployed over the course of the last 10 years alone, Elster and its product sets have gained widespread acceptance by an array of user groups in the flowmetering world.

Within the gas flow market, Elster has focused on upstream, midstream, and downstream gas volume and quality measurement, but targets downstream gas measurement and regulation. Elster Thermal Solutions is focused on all safety, measuring, and controls within the field of industrial heat processes and combustion. Elster sells its gas products to gas and multiutility organizations for use in residential, commercial, and industrial environments.

### Turbine Flowmeter Products

Elster Instromet industrial gas turbine meters and quantometers (short pattern, turbine meters for in-plant measurement) are built for heavy-duty commercial/industrial applications measuring natural gas, air, nitrogen, carbon dioxide, propane vapor, and other noncorrosive gases. They use a turbine wheel technology and display flow volume in actual cubic meters. The company's gas meters fulfill international directives and norms for pressure equipment (PED) and explosion protection (ATEX), as well as metrological requirements and directives.

Honeywell Elster TRZ2 and SM-RI-X turbine meters are approved for custody transfer of natural gas and a variety of technical gases in industrial applications. Series Q and QIC turbine Quantometers are suitable for all noncorrosive gases in a variety of flow ranges, diameters, and pressure ratings. GT/GTS/GTX turbine flowmeters measure natural gas, air, nitrogen, carbon dioxide, propane vapor, and other noncorrosive gases in heavy-duty, large-volume commercial and industrial applications.

## Faure Herman

Faure Herman designs and manufactures helical turbine and ultrasonic flowmeters for custody transfer in the oil and gas industry, as well as turbine and mass meters for the civilian and military aerospace industry. Since 2017, Faure Herman has been part of Le Garrec & Compagnie, a private French group with holdings in the European flow measurement industry. In January 2020, Faure Herman acquired Ultraflux, a fellow French ultrasonic flowmeter company.

### History and Organization

Faure Herman was founded in 1925 by Jean Faure Herman, an automobile importer who realized the need to measure the amount of fuel consumed by aircraft – in those days, attempts to break distance, speed, and altitude records often failed because aviators couldn't tell how much fuel they were using. Herman invented an oscillating piston flowmeter that airlines soon adopted.

Since then, Faure Herman has continued to develop equipment to improve liquid measurement through both mechanical and ultrasonic technologies. The company developed a set of flowmeter technologies used primarily in applications for industrial processes, oil and gas exploration and production, nuclear power, and aerospace.

Faure Herman invented the helical turbine meter in the 1950s as an improvement over the existing flat-bladed technology and pioneered ultrasonic meters in the 1990s with a varying number of transducers (1–18) to optimize customer measurements.

By 1980, the company had expanded internationally and was supplying products for refineries in the Middle East. In 1995, it founded Faure Herman Meter, Inc., in Houston, Texas. In 1999, Faure Herman became part of the French Zodiac aerospace group. In the 2000s, Faure Herman capitalized on its expertise in the high-precision liquids metering market to design and develop a line of ultrasonic flowmeters for transactional and process applications.

In February 2007, the US-based IDEX Corporation acquired Faure Herman from the French investment firm Ciclad. IDEX is an international applied solutions business with a broad line of flowmeters, pumps, valves, and other fluidics systems products. Faure Herman became part of IDEX's Liquid Controls division, which specializes in the development of fluid measurement solutions.

In November 2017, IDEX sold Faure Herman to Le Garrec & Cie, a small family-run company.

On January 17, 2020, Faure Herman acquired 100% of Ultraflux's shares. The two French flow measurement companies had a longstanding professional relationship and decided to combine their expertise and research and development programs.

### Turbine Flowmeter Products

Faure Herman manufactures turbine flowmeters used for both gas and liquid in the oil industry, industrial process control, and aerospace applications. Its Heliflu™ turbine flowmeters use a helical blade technology that makes them insensitive to density and viscosity, allowing them to measure higher viscosities than traditional meters. Faure Herman offers both custody transfer and process control models, in flowrate ranges from 50 L/h to 5,500 $m^3$/h, and for pipe sizes from 1/2″ to 18″.

## HOW THEY WORK

There is an intuitive idea behind the way turbine meters work. Turbine meters have a spinning rotor that is mounted on bearings in a housing. The rotor has propeller-like blades, and it spins as water or other fluid passes over it. The force of the current turns the rotor. Flowrate is proportional to the rotational speed of the rotor. Multiple methods are used to detect the rotor speed. These include an electrical sensor and a mechanical shaft.

Turbine meters differ according to the spinning rotor design. Paddlewheel meters and propeller meters are several designs. Paddlewheel meters are similar to a water wheel and have a rotor that has an axis of rotation that is parallel to the direction of the flow. Many paddlewheel meters are insertion types. Propeller meters typically have only a few blades and have a rotor that is suspended in the flowstream.

What is common to turbine meters is the use of a rotor that spins or rotates in proportion to flowrate. Measuring flow in this way requires the capability of detecting how fast the rotor is turning. For this reason, the turbine meter as it is used today had to wait until the pick-off sensor with a magnet and a rotating conductor were invented. This made it possible to count the number of rotations of a turbine rotor. Turbine meters came to be widely used for the first time during World War II. They were used during this time to measure fuel consumption on military aircraft. Soon after this, they began to be used to measure the flow of hydrocarbons.

The use of turbine meters to measure gas flow dates back to 1953. Rockwell introduced an improved turbine meter to the gas industry in 1963. It took about 10 years for turbine meters to become accepted by the gas industry for measuring gas flow. The American Gas Association (AGA) published its report #7, "Measurement of Fuel Gas by Turbine Meters" in 1981. Since that time, turbine meters have been solidly established in the gas industry, especially for custody transfer applications.

## TYPES OF TURBINE METERS

The following are the main types of turbine flowmeters:

- Axial
- Paddlewheel
- Pelton wheel
- Propeller
- Single jet
- Multi-jet
- Woltman

**Axial** turbine meters have a rotor that revolves around the axis of flow. Most flowmeters for oil measurement and for measuring the flow of industrial liquids and gases are axial flowmeters. Axial meters differ according to the shape of the rotors and the number of blades. Axial meters for gas measurement are designed differently from axial meters for liquid applications. Axial turbines are the most common type of turbine meter.

**Paddlewheel** turbine meters have a lightweight wheel with flat blades that spins in proportion to flowrate. They are used for measurement of low-speed flows. Paddlewheel meters look like a water wheel and are mounted so that the spinning wheel only dips into part of the flowstream. For this reason, many paddlewheel meters are designed for larger line sizes and can be adjusted for insertion depths. Paddlewheel meters are mounted at right angles to the flowstream, and the shaft and bearings are outside of the flow.

**Pelton wheel** turbine meters work somewhat like paddlewheel meters but have a single-size rotor that has straight blades. Pelton wheel meters are based on the original Pelton wheel that was used to turn turbines with flowing water. This original Pelton wheel had bucket-like containers mounted on the wheel that caused it to rotate when water flowed over the wheel. The design of Pelton wheel meters is similar, but they have a more compact design. Pelton wheel meters are used mainly to measure the flow of low-viscosity fluids at low flowrates.

**Propeller** turbine meters are bulk meters used mainly to handle dirty liquids. Propeller meters have helical-shaped blades that are longer than most other turbine meter blades. They also have fewer blades than the rotors of most other turbine meters. Forrest Nagler invented propeller turbines in 1916, and they resemble the wooden propellers found in boats. Because they have few blades, they can more easily handle debris and are among the most rugged turbine meters.

**Jet** turbine meters are primarily used for municipal water measurement, although some are also used for commercial and industrial water measurement. Jet meters are of two types:

- Single jet
- Multi-jet

Single jet and multi-jet meters have one or more orifices that direct a stream or "jet" of water onto a set of blades, causing them to turn. Single jet meters have one orifice, while multi-jet meters have multiple orifices that create streams of water, causing a rotor to revolve.

**Woltman** turbine meters have a rotor whose axis is in line with the direction of flow. Woltman meters are water meters used for larger volume applications. They are also called "bulk" meters and are quite accurate.

There are two other types of meters that are difficult to classify, but they bear mentioning here since they incorporate turbine technology.

**Compound** meters incorporate two-meter technologies. They are a type of hybrid meter. Compound meters are designed to handle both high flowrates and low flowrates. They are often installed in apartment and office buildings that have both periods of high flow use and periods of low flow use. For example, in a large apartment building, flowrates are likely to be very high in the morning and evening. In the day and at night they are much lower.

Compound meters often incorporate turbine technology for high flowrates and PD flow technology for low flowrates. The PD component of a compound meter measures flow during the off hours, because PD meters are very accurate in measuring low flows. During peak usage, the turbine meter does the measuring. Some compound meters use jet-type turbine technology for the low flowrates instead of PD meters, making them a type of dual-turbine technology meter.

**Fire service** meters are used in commercial buildings where there is the need to handle very high flowrates on occasion, such as a situation requiring the use of water from a hydrant to fight a fire. Many fire service meters incorporate turbine technology.

## GROWTH FACTORS FOR THE TURBINE FLOWMETER MARKET

While world economies are recovering from the COVID-19 pandemic, this was followed by a rise in inflation. This was due in part to supply chain issues, as vital parts such as semiconductors became scarce. The surge in pent-up demand for travel, entertainment, and going out to events resulted in demand outpacing supply in many areas. In oil and gas, this has resulted in much higher oil prices, which is likely to work to the benefit of the turbine flowmeter market.

At the same time, Russia's invasion of Ukraine in February 2022 has led to economic uncertainty and instability around the world. Until this situation is resolved or at least downgraded to a low-scale conflict, this uncertainty and instability is likely to be pervasive around the world, especially in the United States and European economies.

Despite these uncertainties, rising energy demand coupled with high oil prices is likely to improve the outlook for turbine flowmeters. In addition, turbine sales are being helped by new applications in oil and gas, cryogenics, ultra-pure water processing, and other applications. Suppliers who support the aerospace industry are reporting an increase in government spending for the military as well as for civilian and government aircraft flight applications.

Growth factors for the overall turbine flowmeter market include the following:

- Installed base of turbine flowmeters
- Turbine flowmeters are used for both liquid and gas measurement
- Approval organizations specify turbine meters
- Turbine flowmeters are a good choice for steady, medium- to high-speed flows
- Turbine suppliers continue to make technology improvements

### Installed Base of Turbine Flowmeters

One major growth factor for turbine flowmeters is their large installed base worldwide. Because they were introduced long before new-technology meters, turbine flowmeters have had much more time to penetrate the major markets in Europe, North America, Asia, and elsewhere.

Installed base is a relevant growth factor because often when ordering flowmeters, especially for replacement purposes, many users prefer to replace like with like. The investment in a flowmeter technology is more than just the cost of the meter itself. It also includes the time and money invested when training people on how to properly install and use a meter. In addition, some companies stock spare parts or even spare meters for immediate replacement purposes. As a result, when companies consider switching from one flowmeter technology to another, there is more than just the purchase price to consider. Thus, the large installed base of turbine flowmeters worldwide will continue to be a source of orders for new and replacement meters in the future.

### Turbine Flowmeters Are Used for Both Liquid and Gas Measurement

Turbine flowmeters are reliable volumetric and mass flow measurement devices for both gases and liquids of any type. This feature is one of the reasons for the technology's continuing popularity as users do not have to necessarily shift from one flowmeter technology to another when the fluid to be measured changes. This competitive advantage is evident in cases where a buyer cannot consider magnetic technology for use with petroleum liquids nor Coriolis technology if the application will include low pressure or low-density gases. Turbine technology is suitable for use in all of these various conditions.

## Approval Organizations Specify Turbine Meters

Turbine meters are specified by approval bodies worldwide for use in custody transfer for utility measurement in residential, commercial, and industrial applications. These organizations include the American Water Works Association (AWWA), the American Gas Association (AGA), the International Standards Organization (ISO) in Europe, and others. These approvals have been in place for many years. Some turbine meters also comply with AGA-8, a more recent certification that deals with the compressibility factors for natural gas and other related hydrocarbon gases.

The approval of a standard by the AGA for using turbine flowmeters for custody transfer of natural gas has been a significant factor in the use of turbine meters for gas applications. However, now turbine meters face competition from ultrasonic meters, and especially since the report on the use of Coriolis flowmeters has also been approved by AGA (Report #11). Differential pressure flowmeters are also widely used for natural gas flow measurement.

So, while turbine meters have enjoyed an advantage in the past based on these approvals, this advantage is diminishing as new-technology flowmeters are approved for custody transfer of natural gas. At the same time, it is likely that turbine flowmeters will continue to be used for gas flow applications because of their price advantages over ultrasonic and even differential pressure flowmeters, as well as due to their continuing technology improvements.

## Turbine Flowmeters Are a Good Choice for Steady, Medium- to High-Speed Flows

Even though turbine flowmeters are losing ground to new-technology flowmeters in some market segments, they remain a viable choice for steady, medium- to high-speed flows. Although turbine meters were invented in the 18th century, they were not widely used in industrial markets until after World War II. Since that time, turbine meters have become solidly entrenched in the oil, gas, water, and industrial liquid flow measurement markets.

Turbine meters excel at measuring clean, steady, medium- to high-speed flows of low-viscosity fluids. This characteristic contrasts with another conventional metering technology, PD flowmeters, that excel at measuring low-speed flows and high-viscosity fluids. Thus, while there are some applications in which these two standbys do compete, turbine and PD flowmeters are more complementary than competing. It is also true that PD meters do better in the lower line sizes, whereas turbine meters excel in the larger line sizes.

### Turbine Suppliers Continue to Make Technology Improvements

Turbine meter suppliers have made technology improvements to make turbine meters more reliable. Many of these improvements have involved making the moving parts – a traditional source of concern regarding maintenance and repair – more reliable. By making the ball bearings out of more durable materials such as newly developed ceramics and synthetic sapphires, turbine suppliers have been able to add significantly to the life of the bearings. This is important since some customers select new-technology meters over turbine meters simply because turbine meters have moving parts that are subject to wear.

There are now turbine designs that meet sanitary guidelines. When used with sanitary connections, turbine meters control flow in food and beverage applications. These meters are not the best choice for low-flow applications, however, they do provide high accuracy. Irrigation and water purification are two other common applications for turbine flowmeters. Here, the turbine's inherent ruggedness is a desirable design feature.

Other product enhancements that are available today include the dual-rotor design being promoted by Cox Flow Measurement (now part of Badger Meter) and other manufacturers. The dual-rotor design increases the effective operating range of turbine meters in the smaller line sizes. This innovation specifies that the two rotors turn in opposite directions, with the first rotor being upstream from the second and acting as a flow conditioner. Flow is then directed back to the second rotor. The rotors are hydraulically connected and will continue to turn as the flow decreases even at very low flowrates. This innovation has enhanced turbine flowmeters' suitability in low-flow applications (Figure 7.1).

The use of robust materials in construction has created a class of turbine flowmeters rated as "industrial grade" by some manufacturers (e.g., Cameron). This class of meter often has improved accuracy and linearity. Bi-directional turbine flowmeters are now also available and have found use in several industries such as Oil & Gas and Water & Wastewater. And suppliers have introduced battery-powered turbine flowmeters into the market (e.g., Universal Flow Monitors) with battery lives of 2 years or more, making models so equipped potentially a more economical choice.

## FACTORS LIMITING THE GROWTH OF TURBINE FLOWMETERS

While there are many positive growth factors for turbine flowmeters, there are also some limiting factors. Some suppliers cite a saturated market and heavy competition, especially from new-technology flowmeters. Increasingly, applications are requiring higher accuracy and turbine flowmeters – despite their design improvements – are perceived as a less accurate technology. Suppliers also report difficulty competing against magnetic and ultrasonic meters in the larger line sizes, and turbine meters cannot offer the situational versatility of being offered in a clamp-on configuration.

**FIGURE 7.1** An industrial gas turbine flowmeter.

Below are the other selected factors limiting the growth of the turbine flowmeter market:

- Maintenance and repair
- Competition from new-technology flowmeters
- Competition from conventional flowmeters
- A perceived lack of investment in enhanced turbine flowmeter designs
- Turbine flowmeter market not keeping pace with other flowmeter types

## Maintenance and Repair

Suppliers cite the maintenance and repair regimens required to maintain accuracy as perhaps the biggest barrier to growth in this market. Depending on the condition of the fluid, a rotor will eventually need to be replaced to ensure measurement accuracy. Other factors cited include learning how to use and integrate turbine meters in larger control network applications and the fact that turbine meters are not a viable choice for steam flow management.

## Competition from New-Technology Flowmeters

Turbine flowmeters face stiff competition from new-technology flowmeters, especially Coriolis and ultrasonic meters. Both the Coriolis and ultrasonic flowmeter markets are growing at a substantial rate. This trend is expected to continue.

Turbine flowmeters face competition from new-technology meters particularly in the areas of oil and gas measurement. One major competitor to turbine meters for oil measurement is Coriolis meters.

Coriolis meters have been developed for use in trucking and transportation measurements, and also for in-plant measurement of hydrocarbons. While Coriolis meters are typically more expensive than turbine meters, some users are willing to pay the difference in cost because of the high value of the product being measured, the reduced demands for maintenance that is promised by the nonintrusive Coriolis design, and the inherent capability to generate a mass flowrate.

Turbine meters also face competition from ultrasonic meters for custody transfer of natural gas, especially in the larger pipe sizes (12 inches and above). Ultrasonic meters excel in larger pipe sizes, and they were approved in 1998 by the American Gas Association for use in the custody transfer of natural gas. Coriolis flowmeters, by contrast, become unwieldy and expensive in sizes above four inches. While turbine meters are losing ground to ultrasonic meters for natural gas flow measurement, they still retain a price advantage over ultrasonic meters. Some companies may also hesitate to make the switch to ultrasonic meters if the technology is unfamiliar to them. For these reasons, a substantial number of turbine meters will still be sold for custody transfer of natural gas.

Over time, magnetic flowmeters will begin to have more of an impact on the sale of turbine flowmeters for utility measurements. The American Water Works Association (AWWA) has approved a standard for the use of transit time ultrasonic flowmeters for water supply service applications. The AWWA has also approved the use of magnetic flowmeters for custody transfer of water. The organization has also published information on the use of Coriolis flowmeters in revenue-producing applications. Turbine meters are facing continuing competition from new-technology flowmeters for water utility measurement, and this competition may well become even more intense over time.

## Competition from Conventional Meters

Turbine meters also face competition from PD and from compound meters. PD meters are widely used for custody transfer of water, as are compound meters. PD meters are also used for measurement of industrial liquids and gases. However, in some ways, these markets are more complementary than competing. Turbine meters perform well at steady, higher flowrates. PD meters perform extremely well at low flowrates. This attribute explains why PD meters are widely used to measure water for custody transfer in homes and in commercial buildings. Flowrates in these buildings can be very low or nonexistent when no one is at home or in the buildings. So even though PD meters do compete with turbine and compound meters, they are best used for applications where turbine meters do not do as well. These applications include low flowrates and highly viscous liquids.

## A Perceived Lack of Investment in Enhanced Turbine Flowmeter Designs

Many of the larger flowmeter companies have chosen not to invest in turbine flowmeters, including companies such as KROHNE, Yokogawa, and Endress +Hauser. On the other hand, several important flow companies have extensive turbine flowmeter product lines. Daniel (now part of Turnspire Capital Partners) has several turbine meter lines, and Elster is also a major supplier of turbine meters while growing the presence of their ultrasonic line at the same time.

Some companies may have chosen not to invest in turbine technology because they see it as older technology, or because turbine meters have the unacceptable "moving parts". However, some segments of the turbine market are actually growing, such as the municipal gas and oil application segments. Despite these facts, there is less overall investment going on in new turbine meters than in new-technology meters. This is a negative growth factor for the turbine flowmeter market.

On the other hand, the large number of suppliers will ensure that new products continue to be developed. For example, Hoffer improved its turbine meters with designs specifying hybrid ceramic ball bearings and unique blade angles. So even though there is less new product development going on for turbine meters than for new technology meters, individual innovations from the substantial number of turbine suppliers will help to make up for this perceived inadequacy.

## Turbine Flowmeter Market Not Keeping Pace with Other Flowmeter Types

Some people believe that the turbine flowmeter market is declining, and say that turbine meters are being replaced by Coriolis, ultrasonic, and magnetic flowmeters. The fact is that the turbine flowmeter market has shown only slow growth for the past 5 years. So, the good news is that the market is not declining, but the bad news is that during this same time period the Coriolis, ultrasonic, and magnetic flowmeter markets have expanded substantially. So, even though the turbine market may not be declining, it is not keeping pace with new-technology meters.

Part of the issue with turbine meters is research and development. There is a group of suppliers, including Elster, Badger Meter, and Hoffer, among others, that are actively researching this market and introducing new features and products. Examples include different bearing types, different rotor designs, dual rotors, and bi-directional capability. Other suppliers continue to offer products but are not actively engaging in new product development. This prevents the market as a whole from expanding significantly.

A major reason for the expansion of the Coriolis, ultrasonic, and magnetic flowmeter markets is the large amount of research and development that translates into new products. In addition to this, turbine meters have moving parts and require periodic maintenance, while many end-users are selecting meters with no moving parts and minimal maintenance. They are also attracted by the high accuracy of new-technology meters. These are the reasons why the new-technology meter markets are expanding so rapidly. There is still hope for turbine meters to expand, but this requires a broader commitment by suppliers to offer improved products.

## APPLICATIONS FOR TURBINE METERS

**Utility Measurement:** PD meters dominate the residential flowmeter market, and PD diaphragm meters have a foothold in the utility/billing area. Single jet and multi-jet turbine meters are used to measure water flow in private homes. Turbine meters are used to measure water flow in hotels, office buildings, apartment complexes, and other commercial buildings. These buildings are most likely to use single jet or multi-jet meters. They may also use Woltman or compound meters, depending on the volume of flow.

For utility measurement, single jet, multi-jet, compound, and Woltman turbine meters compete more against PD meters than new-technology meters. This is because industry approvals for utility and billing by new-technology meters such as Coriolis, vortex, magnetic, and ultrasonic are still undergoing development and acceptance.

**Submetering:** The increased cost of water and the increased size of apartment and office buildings has generated the practice of submetering. Submetering typically takes place in an office or large apartment building with many tenants. Under submetering, the tenants are metered individually for their water use. This practice works just as well for offices as for condos or apartments.

By using submetering, the building owner can pass along the costs of water use to the tenants, based on their actual usage. Some companies specialize in submetering, and often turbine meters are used for this purpose. These meters are typically single jet or multi-jet meters. Submetering also permits a company to better manage its overall energy conservation efforts, as the firm can acquire more local data and pinpoint specific energy sinks or wasteful practices.

**Industrial Use:** Turbine meters are used to measure water use at commercial businesses and at industrial plants. This is a utility measurement; however, since it occurs at a manufacturing plant it is considered an industrial measurement. Other types of meters used for this purpose include vortex, magnetic, and DP.

**Comparison to PD Meters:** Turbine flowmeters are used when the line sizes are larger and the flow volume is greater than can be handled by PD meters. PD meters do better with low flowrates and low flow volumes, while turbine meters excel with medium to high flowrates and flow volumes. Therefore, turbine meters are often used in larger line sizes, especially those greater than four inches.

**Oil:** There is a very large market for measuring the flow of oil, refined fuels, and hydrocarbon products. Some suppliers are focused on providing meters for loading and unloading trucks, tankers, ships, and airplanes. Most of this measurement is for custody transfer purposes, and it occurs both upstream and downstream of refineries. Oil trucks that deliver oil to businesses, ships, and aircraft use a flowmeter to measure how much oil is dispensed. Both turbine meters and PD meters are used for this purpose. PD meters are used for flow measurement of heavier, higher viscosity oils, while turbine meters are used for the lighter, less viscous oils. Coriolis meters are also now increasingly being used for these applications.

**Process Liquids:** Besides water and oil, turbine meters also are used to measure process liquids. These include pharmaceutical chemicals, paints and varnishes, industrial chemicals, dairy products, printing ink, cosmetics, and many other liquids. In many cases, turbine meters provide a highly accurate measurement but are less expensive than other meters such as Coriolis or magnetic. Turbine meters are used for higher flowrate measurements, while PD meters are used for measurement of flow at lower flowrates. Coriolis meters are sometimes used when high accuracy and long-term reliability are important measurement criteria.

**Gas:** Turbine meters are used as a billing meter to measure the amount of gas used at commercial buildings and industrial plants. They are also used in custody transfer and noncustody transfer applications of natural gas in upstream and downstream production environments. Examples of commercial buildings that use gas flowmeters for billing are apartment buildings, hotels, and office complexes. Industrial plants such as food processing, chemical, and pharmaceutical plants also use turbine meters for billing purposes. These meters are different from the meters used to measure gas as part of the manufacturing process.

The market for commercial and industrial turbine gas meters for billing purposes or to track gas usage at a commercial or industrial building is known as the municipal gas market. This is a utility measurement and a specific group of companies supplies meters to this market.

Turbine flowmeters are also widely used for the custody transfer of natural gas. They measure gas flow on large gas pipelines that carry natural gas from its source to its destination – in some cases for thousands of miles. The main competitors to turbine meters for this important application are ultrasonic and DP flowmeters.

## FRONTIERS OF RESEARCH

The following are frontiers of research for turbine flowmeters.

### Improved Construction Materials for Bearings

One way in which turbine flowmeters are subject to wear is due to their moving parts. Suppliers have reduced their susceptibility to wear significantly by making their bearings out of much stronger materials than the traditional steel bearings. These include stainless steel, sapphire, alumina oxide ceramic, silicon nitride ceramic, and heavy-duty plastics. Ceramic has a porous, glass-like surface so that it is virtually frictionless and needs very little or no lubrication to operate. These advanced bearing materials extend bearing life, and hence the life of the turbine flowmeter.

Another way to improve on bearing design is through different bearings styles. These include radial deep groove bearings, thrust bearings, and angular contact bearings. Bearings can also be designed to fit particular applications by the use of special lubrication and coatings.

## Dual Rotor Designs Enhance Performance

Cox Instrument, now part of Badger Meter, has introduced dual rotor turbine flowmeters. With the dual rotor design, two closely coupled rotors turn in opposite directions. The average of the two rotors is used to determine flowrate. This design lessens the effect of swirl in the flowstream, and greatly reduces the need for flow conditioners and upstream and downstream straight pipe runs. This design also significantly increases the flow range of the meter. This greater turndown makes it possible to use only one turbine meter to measure flow with both high speed and lower speed flow when normally more than one flowmeter would be required.

Combining the dual rotor design with other technological advances can further enhance the performance of dual rotor meters. This includes ceramic bearings, helical rotors, bearing diagnostics, and advanced flow processors. These enhancements, together with the dual rotor design, open up new applications for turbine meters, and improve the performance and life expectancy of these turbine meters.

## Dual Tube Design

The author has two patents on a dual tube meter, which includes a turbine design. Rather than the dual rotor design, the dual tube design has two rotors side by side inside the meter body. Each rotor has its own pickup coal, and the two rotors operate independently. The outputs from both rotors are relayed to a transmitter. The transmitter compares the two outputs from the rotors and creates a single measurement reading. This single reading is based on a previous calibration of the meter (Figure 7.2).

Discussions by Flow Research with engineers who use turbine meters to measure the flow of refined fuels onto tankers found that occasionally a solid impurity in the flowstream impacts the turbine blade and interferes with a correct reading. This may not be discovered until the meter is calibrated again. With the dual tube meter, even if one rotor is knocked out of calibration, the other rotor will continue to read the flowrate. The transmitter can be programmed to set an alarm condition when the readings from the two rotors vary more than a specified amount. With the alarm set, an operator can be notified and the affected rotor can be repaired.

Another application has to do with environmental reporting. For some applications, the Environmental Protection Agency (EPA) requires 24/7 reporting of flowmeter readings. If a turbine meter is being used and its rotor is knocked out of calibration, the reporting may be incorrect, or it may cease altogether. With the dual tube meter, reporting continues from at least one rotor, and the other rotor can be repaired if an alarm condition is set.

One advantage of the dual tube meter is that it can accommodate different types of sensors inside the meter body. A prototype of a turbine design has already been

**FIGURE 7.2** Inside view of patented Yoder Dual Tube turbine meter showing the rotors inside the meter body – Photo courtesy of Jesse Yoder.

built, as well as a prototype of a vortex meter design. Both prototypes have been tested at the Colorado Experiment Engineering Station, Inc. (CEESI). It is also possible to incorporate two different types of sensors inside the meter body, thereby building a compound meter that can accommodate flows of different velocities or properties.

## Other Improvements

In addition to different materials of construction and rotor design, manufacturers are adding advanced diagnostics to improve the performance of turbine meters. Turbine meters are widely used to measure the flow of natural gas, and they are used for custody transfer of natural gas. This is a highly sophisticated measurement that requires high accuracy. Honeywell Elster is one of the leading turbine suppliers of turbine meters for these applications. Other flowmeters that compete in this application include ultrasonic and differential pressure flowmeters.

# 8 Open Channel Flowmeters

## OVERVIEW

So far this book and the preceding volume have discussed different methods of measuring flow in a closed pipe. In fact, most of the flowmeters discussed, with the exception of magnetic flowmeters, require that the flow being measured occur in a closed pipe. However, when considering all the water present on the earth, and the many rivers and streams in which it flows, it might seem that measuring open channel flow should be a much more important topic than measuring flow in closed pipes.

The answer to this question is that in most cases where the flow has to be measured, the fluid is in a pipe or in some kind of physical container. Also, it is not only water whose flow is measured. Oil, gas, chemicals, pulp, slurries, and refined products need to be measured, especially when they are bought or sold. Water is measured intensively in municipal water and wastewater industry, but there it is usually in pipes. On the other hand, at times we do need to measure the flow of water in rivers and streams. This is especially true today in drought-stricken areas like parts of California, where water is becoming an increasingly important though increasingly scarce commodity.

The Quabbin Reservoir, along with the Ware River, Wachusett Reservoir, and contributing tributaries in central Massachusetts are the source of clean water for 51 Greater Boston communities. It travels mostly in closed pipes to the Boston area, where it is sent to various water utility companies that distribute the water throughout Boston. While the bulk of this flow is in closed pipe, open channel flow occurs at the beginning of the reservoir and at several points along the route to Boston.

Municipal water and wastewater treatment plants provide another example of mixed open channel and closed pipe flow. Most of the flow inside the plants is in closed pipes. But open channel flow may occur as rivers and streams serve as sources for the water that is processed in a municipal water plant for drinking. Some water also comes from wells.

Once households and businesses use the water, it needs to be processed in a wastewater treatment plant. Used water is treated for reuse, recycling, or to be discarded. The output from some wastewater plants is turned into fertilizer. There may be open channel flow leaving a sewage or wastewater treatment plant, possibly containing slurries depending on what needs to be done with the output from the plant. While much of the flow inside the plant occurs in closed pipes, the flow from the plant may be in open channels or in partially filled pipes. Once it leaves the

DOI: 10.1201/9781003130024-8

plant, it moves through closed pipes for sanitary purposes and is distributed to different communities for drinking and other household purposes.

The above examples show that closed pipe and open channel flows coexist in many cases and that both may be needed to achieve certain essential purposes. This is why open channel flowmeters are studied alongside closed pipe flowmeters – they are both needed and in some cases are used in the same location or facility for different purposes.

While it is clear that both closed pipe and open channel flows are important and need to be measured, closed pipe flowmeters vastly outnumber open channel flowmeters. Furthermore, there are significantly more technologies used for closed pipe measurement than for open channel measurement. Why is this? While this question can be answered, it is worthwhile to first look at the amount of water there is, and to what extent it needs to be measured.

## AN OCEAN OF WATER

Scientists estimate that about 71% of the earth is covered with water. Of this amount, 96.5% exists in the oceans, which is salt water. This means that 3.5% of the water on the earth is fresh water. This is the water that fills our rivers, lakes, streams, canals, and other inland waterways. It is fresh water that is used for drinking, washing clothes, bathing, and many other household tasks. Businesses equally rely on fresh water for similar purposes. Salt water is used for recreational purposes, and for the transportation of boats and ships. The ocean is also a vast source of food for people and animals, and plays a critical role in the earth's weather systems.

## WATER AND AIR CURRENTS

While most of this chapter is about measuring freshwater flow, it is interesting to think about flow in relation to the ocean. At first thought, it might seem that the ocean doesn't flow, although it laps on the shore. But there are massive ocean currents that flow within the ocean. Ocean currents are the predictable, continuous, directional movement of seawater, driven by wind, gravity, and water density. This is considered horizontal motion. Water also moves vertically when it becomes vapor and rises from the oceans to form clouds. This is the process of evaporation when a liquid becomes a gas. This process is shown in Figure 1 of chapter 4 of Volume I in this series. This figure shows warm air rising to form clouds, which in cold temperatures are made up of ice crystals. The ice crystals collide, grow together, and form snowflakes, which fall to the ground. In warmer weather, the warm air rises to form water droplets, which under the right conditions fall as rain.

It might seem that with all that water around, meteorologists would be interested in open channel flowmeters. However, they typically use satellite imagery, including infrared sensors, to track the flow of ocean currents. One way to judge the speed of water currents is to throw a light object into the current and calculate the time it takes to travel a known distance. This is like the area velocity method of measuring flowrate in that this method includes measuring the speed or velocity of the flow. Another way that oceanographers use open channel-related equipment is by using ultrasonic waves to judge the depth of the ocean.

Water currents are not the only types of currents that meteorologists study. Air currents play a major role in our weather and are concentrated areas of wind. Typically, air currents involve the movement of air from one location to another due to differences in atmospheric temperature or pressure. Some of this movement is caused by the heat from the sun. As the sun heats the ground, it does so unevenly. As the ground is warmed, the air above it rises. This leaves an area of cooler air beneath the rising warm air. The surrounding air moves in to replace the cooler air, creating wind. Air generally moves from high-pressure areas to low-pressure areas, and this is what causes wind.

Measuring air velocity has played an important role in the development of flow instrumentation. Most people are familiar with anemometers that are used to determine wind speed. Sometimes, meteorologists hold them up when giving forecasts "on location" to demonstrate the speed of the wind (Figure 8.1).

In Chapter 8 of Volume I, we traced the development of thermal flowmeters from the studies of hot-wire anemometers at TSI in Minnesota, which were being used to measure air currents in a room using a probe. One line of thermal flowmeters was developed by creating an industrially hardened version of these hot-wire anemometers. This development resulted in the founding of Kurz Instruments and Sierra Instruments in Monterey, California, in 1976.

Open channel flow occurs when liquid flows in a conduit, canal, or waterway with a free surface. Rivers, canals, streams, and irrigation ditches are examples of open channels. There are many occasions when open channel flow needs to be

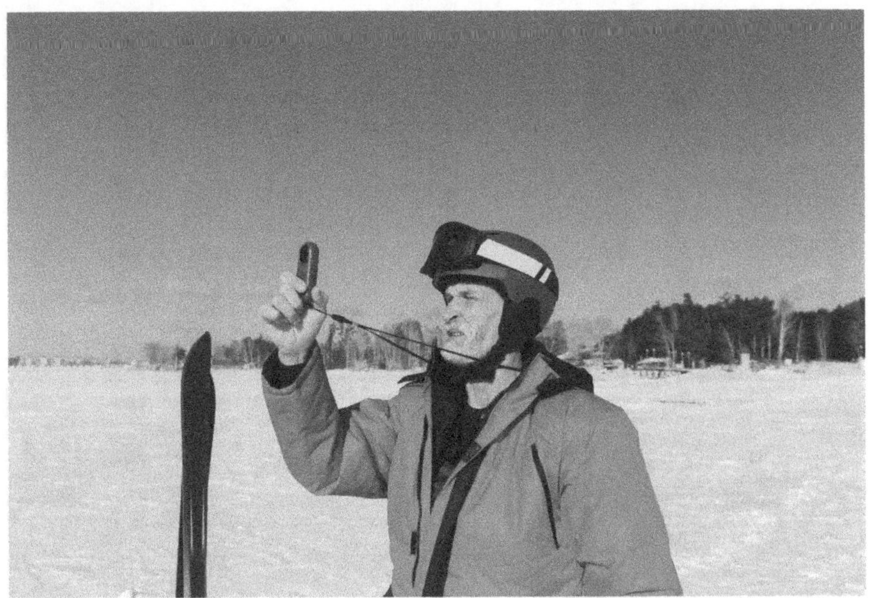

**FIGURE 8.1** Meteorologist holding an anemometer.

measured. Many towns and municipalities that monitor drinking water and wastewater treatment flow use open channel measurement. Open channel flow is also measured in industrial wastewater treatment plants.

## OPEN CHANNEL FLOW VS. CLOSED PIPE FLOW

What may seem confusing about the term "open channel" is that the flow of liquids in partially filled pipes – when not under pressure – is considered open channel flow. Water flowing through a culvert located underneath a street is considered open channel flow, for example, even though the channel is not actually "open." Likewise, flow in sewers and tunnels is classified as open channel flow, along with other closed channels that flow partially filled (Figure 8.2).

One way to understand the difference between open channel and closed pipe flow is to think of it as the difference between gravity-induced and pressurized flow:

- Flow in uncovered channels such as irrigation ditches depends on gravity. Likewise, flow in partially filled closed conduits, such as culverts and drainpipes, also is gravity dependent.
- Flow in closed pipes for industrial applications occurs under pressure.

So open channel flow might be called gravitational flow, while closed pipe flow could be called pressurized flow. This explains why flows in both uncovered conduits and partially filled pipes are considered to be open channel flow: They are both examples of gravitational flow.

**FIGURE 8.2** A flowing stream near Boulder, Colorado – Photo courtesy of Jesse Yoder.

## OPEN CHANNEL FLOWMETER COMPANIES

### HACH

Hach, a Danaher company, manufactures and distributes analytical instruments and reagents used to test water quality and other aqueous solutions. Hach is a world-class manufacturer of open channel flowmeters. In addition to its US headquarters, the company maintains offices in Canada, China, Germany, Australia, New Zealand, India, and Singapore. Hach's flow products and services include legacy brands Sigma and Marsh-McBirney. The company offers wireless flow loggers and noncontact sensors as well as data delivery services. Hach's analytical instruments, services, software, and reagents are used to ensure the quality of water in a variety of industries in more than 100 countries globally.

#### History and Organization

Hach traces its roots to 1933 when Dr. Bruno Lange GmbH founded Lange in Berlin, Germany, to give water quality professionals confidence in their analysis. In 1947, Clifford and Kathryn Hach started the Hach Company in Ames, Iowa. The couple developed a simplified titration method for measuring hardness in drinking water. The Hach Europe office was opened in 1972 in Namur, Belgium. In 1999, the European facility moved to the Düsseldorf location of its sister company, Dr. Bruno Lange GmbH, offering increased capabilities as a distribution and support center for customers in Europe and Mediterranean Africa.

In 1999, Hach became a wholly owned subsidiary of Danaher Corporation. That same year, Danaher also acquired American Sigma, which became part of the Hach Company. American Sigma, founded in 1980 in Western New York, was a worldwide leader in the design and manufacture of flow, sampling, rain, and water quality instruments, communication products, and data management software. American Sigma had acquired its line of closed channel flowmeters from Exidyne Instrumentation Technologies (EIT), which at the time was owned by Bacharach, Inc.

In March 2006, Hach acquired Marsh-McBirney of Frederick, Maryland, a 35-year-old maker of flow measurement equipment for water and wastewater system applications for both the industrial and municipal markets.

#### Open Channel Flowmeter Products

Hach flowmeters offer open channel options for stationary, portable, permanent, and temporary usage, and include extremely rugged portable loggers. Hach's handheld FH950 ultrasonic meter is designed for wastewater and environmental flow monitoring.

The FL900AV system is based upon Doppler technology and is designed to use the Sigma Submerged AV Sensor as its measuring point. The FL1500 is a stationary flow monitoring solution for measuring and logging open channel flow and is compatible with all of Hach's flow sensor technologies including noncontact radar, magnetic, depth pressure, Doppler, ultrasonic, and bubble level.

## Siemens

Siemens, among the largest companies in the world, is a manufacturer of pressure, level, and temperature transmitters, flowmeters, programmable logic controllers, drives, and many other automation products. The company is a worldwide market leader in automation systems and supporting technologies, including process analytics, weighing and batching, measuring, and condition monitoring.

The German company operates throughout 190 countries and has a large global partner network. On October 1, 2020, after spinning off part of the company, Siemens began operating with four industrial businesses – Digital Industries, Smart Infrastructure, Mobility, and Siemens Healthineers – plus Siemens Advanta, Siemens Financial Services, and Portfolio Companies. Digital Industries, based in Nuremberg, offers products, system solutions, and services for automation in the discrete and process areas, including flowmeters and pressure transmitters.

### History and Organization

The history of Siemens goes back to 1847 when the Siemens & Halske Telegraph Company was founded in Berlin. Wilhelm Siemens took over this company in 1850. In 1897, Siemens & Halske became Siemens AG, a stock corporation.

In February 2000, Siemens Energy and Automation and Moore Process Automation Solutions announced a merger. Siemens also purchased Milltronics, a level instrumentation supplier based in Canada and the United States. Siemens later also bought Applied Automation.

In September 2003, Siemens expanded within the international process automation market by acquiring Danfoss' flowmeter division, including its Coriolis, magnetic, and ultrasonic flowmeters. Digital Industries' process automation business now manages this flowmeter line.

### Open Channel Flowmeter Products

Siemens' open channel ultrasonic products include the SITRANS LUT and the MultiRanger/HydroRanger, a versatile short- to medium-range ultrasonic single- and multi-vessel level monitor/controller for virtually any application in a wide range of industries. Overall, key applications for this group include wet wells, flumes/weirs, bar screen control, hoppers, chemical storage, liquid storage, crusher bins, and dry solids storage.

## HOW THEY WORK

Open channel flowmeters are the only game in town when it comes to open channel applications. However, there are different types of open channel meters. The two main types are as follows:

- Hydraulic Structure – Weirs and Flumes
- Area Velocity

**Weirs and Flumes:** A common method of measuring open channel flow involves the use of a hydraulic structure such as a weir or flume. These hydraulic structures are called primary devices. A primary device is a restriction placed in an open channel that has a known depth-to-flow relationship. In that sense, a primary device resembles a primary element placed in the flowstream to create a constriction in the flow to enable differential pressure measurement. Once a weir or flume is installed, the flowrate can be calculated from a measurement of the depth of the water. Charts are available that correlate various water depths with flowrates, taking into account different sizes and types of weirs and flumes (Figure 8.3).

A weir resembles a dam placed across an open channel. It is positioned so the liquid can flow over it. Weirs are distinguished by the shapes of their openings. Types of weirs include

- V-Notch
- Rectangular
- Trapezoidal

Water depth is measured at a specific place upstream from the weir. Each type of weir has an associated equation for determining flowrate (Figure 8.4).

A flume is a specially shaped portion of the open channel with a different area or slope from the channel's area or slope. The velocity of the liquid increases and its level rises as it passes through the flume. Liquid depth is measured at specified points in the flume to determine flowrate. An equation is associated with each kind of flume, taking into account flume size.

**FIGURE 8.3** Flume of an old gristmill.

**FIGURE 8.4** Water flowing over a weir.

**Area Velocity:** It is possible to measure flow without a hydraulic structure such as a weir or flume. In the area-velocity method, the mean velocity of the flow is calculated at a cross-section, and this value is multiplied by the flow area. Normally, this method requires two measurements: mean velocity and the depth of the flow. The area-velocity method is often used when it is impractical to install a weir or flume, and for temporary flow measurements. Examples include sewer flow monitoring and influx and infiltration studies. In the area-velocity method, flowrate $Q$ is calculated using the following continuity equation:

$$Q = A \times v$$

Important driving forces behind the open channel flowmeter market are an increasing need to measure water flow due to population growth and environmental regulations. Unusual weather patterns that create droughts and dry water beds are becoming more common, both in the United States and in other regions of the world. Just as is the case with oil and gas, water needs to be measured when ownership is transferred from one person or entity to another. Custody transfer of water is a critical measurement, just as is custody transfer of oil and gas, even though the transfer of water ownership may sometimes be treated as a billing or utility measurement.

There are three other methods of open channel measurement that are less common than hydraulic structures or area velocity, but are still worth describing:

- Dilution
- Timed-Gravimetric
- Manning Formula

**Dilution:** A tracer such as a fluorescent dye or radioactive iodine is added to the flow. Downstream, the concentration of the tracer is measured and this value is used to compute flowrate, based on a theoretical formula. Two techniques used are the constant-rate injection method and the total-recovery method.

**Timed-Gravimetric:** This is a kind of sampling method for measuring open channel flow. The liquid is captured in a container for a specific period of time. It is then weighed using a force or mass measuring device such as a load cell or beam scale. A similar method uses a container having a known volume to capture the flow. Using a stopwatch, flow is calculated based on how much liquid was captured in the measured time. This method can yield high accuracy but does not lend itself to continuous measurement.

**Manning Formula:** This method was proposed in 1889 by an Irish civil engineer named Robert Manning. The method was modified in the 1930s. The Manning formula bases its calculation on the following three variables:

- The slope of the water surface
- Cross-sectional area
- The roughness of the conduit

This method is less accurate than the area-velocity method because it uses assumed values rather than measured values.

For more information on all these methods, refer to the excellent reference book called *Teledyne Isco Open Channel Flow Measurement Handbook.*

## GROWTH FACTORS FOR OPEN CHANNEL FLOWMETERS

The future of open channel flow measurement has oftentimes been uncertain. Open channel measurement declined somewhat with construction of nuclear power plants, for instance, but its future there may rebound as the perceived dangers of nuclear power are outweighed by their reduced effects on the environment in generating electricity. Similarly, users have been slow to invest in open flow measurement much beyond its use in municipal systems, and then only within long-term engineering cycles. But the dollar value of water and the need for its re-use have introduced a new set of applications for this measurement. The selected growth factors for open channel measurement are as follows:

- An increasing need to measure water flow.
- Conventional flowmeters are being replaced by new-technology flowmeters.
- Flowmeters are used at multiple places in an irrigation system.
- Increased regulations and cost drive flowmeter usage.
- Desalination plant construction is a boon to instrumentation suppliers.
- Continuing development of technologies.

## An Increasing Need to Measure Water Flow

An increasing need to measure water flow due to population growth and environmental regulations are important driving forces behind the open channel flowmeter market. While energy needs capture many of the headlines, the fundamental need to supply clean water to many populations, and also to provide wastewater services, is becoming more important. Besides increased population, unusual weather patterns that create droughts and dry water beds are becoming more common. This is true throughout the world whether the location is sub-Saharan Africa, Western United States, or the Middle East. The monitoring and control of open channel water sources has assumed much importance in the last several years and this trend will continue into the foreseeable future.

## Conventional Meters Are Being Replaced by New-Technology Flowmeters

Magnetic flowmeters and ultrasonic flowmeters are displacing conventional flowmeters such as differential pressure (DP) and turbine flowmeters in some agriculture/irrigation applications. Magnetic flowmeters have significant advantages over conventional flowmeters. Specific to turbine types, they are mainly displacing propeller and paddlewheel flowmeters. And regarding DP meters, and most particularly orifice plate meters, magnetic meters do not have a primary element susceptible to wear that can significantly degrade measurement accuracy. With no primary element to replace, and no moving parts to introduce wear, magnetic flowmeters represent a very stable and reliable long-term method of measurement with minimal maintenance costs. For these reasons and others, magnetic flowmeters are displacing conventional flowmeters for some agriculture/irrigation applications.

The advantages of magnetic flowmeters are especially relevant when measuring water from a water source. Flowmeters are often positioned in the flowstream right after a pump that is pumping water from a well or other water source. If this is clean water, a propeller or paddlewheel meter may work quite well. However, if it has particles or debris in it, this will flow through a magnetic flowmeter without problem, while it might get caught up in a propeller or paddlewheel meter. In some cases partially full magnetic flowmeters can replace area-velocity methods of measurement.

The same comment applies to ultrasonic flowmeters. Both magnetic and ultrasonic flowmeters have no obstruction in the flowstream, while a turbine flowmeter determines flowrate from the speed of rotation of rotors located in the flowstream. As these rotors are in the flowstream, they necessarily intrude into it, making them vulnerable to flowstream impurities or debris.

## Flowmeters Are Used at Multiple Places in an Irrigation System

Flowmeters are used for measuring input water, the amount of water output from an irrigation system, chemical feed, and for batching purposes. Because there are multiple uses for flowmeters in irrigation, there are multiple opportunities for flowmeter use. Some decisions about the type of flowmeters used are based in part on line size. Some water input lines can be 12 inches or larger, though their diameter varies with the type of irrigation system. A larger line size may favor magnetic or ultrasonic meters, both for accuracy reasons and for speed of throughput.

Ultrasonic flowmeters have an advantage in that they can be mounted inline and also in a clamp-on form. Clamp-on ultrasonic meters can be used to check the accuracy of other meters. Because they are in a clamp-on form, they can be moved around from one location to another in case multiple meters need to be checked or verified. Transit time ultrasonic meters can accurately measure clean water and water with some impurities. However, ultrasonic Doppler meters can accurately measure dirty water and water with particles or impurities. As a result, both transit time and Doppler ultrasonic meters are used in irrigation systems. Irrigation systems are likely to include a combination of closed pipe and open channel flowmeters.

## Increased Regulations and Cost Drive Flowmeter Usage

An increased number of state and federal regulations require tracking water usage in agricultural and irrigation applications. Water is becoming scarcer and a more precious resource. This includes surface water in lakes, rivers, and streams, along with groundwater in wells and aquifers. As a result, end-users are moving towards more accurate flowmeters, some of which contain totalizing capability. Users are also increasingly demanding improved totalization features as a way to assess present and future irrigation requirements with an eye toward economy and resource conservation. They are also using flowmeters with protocols that can communicate with other instrumentation. This capability facilitates more efficient measurement and control within irrigation systems.

When irrigation began to be modernized in the 1930s and 1940s, DP flowmeters were pretty much the only game in town. Both Venturis and orifice plates were used in irrigation systems. Some end-users may still select them today as a new meter purchase, especially because DP transmitters with Venturi meters offer minimal obstruction even if they do cause pressure drop. There is also a replacement market for the purchase of DP flowmeters as they can potentially provide many years of service. However, both magnetic and ultrasonic meters provide better accuracy than DP meters, so some DP flowmeters still in service are being replaced by these two types of new-technology flowmeters.

## Desalination Plant Construction Is a Boon to Instrumentation Suppliers

As climate change intensifies and water becomes a scarcer commodity, one answer for both irrigation and drinking water can be turning seawater and brackish water into fresh water through desalination. Presently, desalination is a long-term solution, and a strong growth area for flowmeter manufacturers presently targeting agriculture and irrigation. Desalination has grown steadily in the last decade. More than 300 million people around the world now get their water from approximately 20,000 desalination plants.

Challenges still exist – primarily cost and environmental concerns – but they are also being mitigated to some degree by technology. In the last three decades, the cost of desalination has been reduced dramatically. On the environmental end, renewable sources are now able to provide some of the large amounts of energy required, and some efforts are underway to cope with serious concerns about harm to marine life from the plant intake systems and their briny wastewater.

Saudi Arabia leads the world in freshwater production from seawater, with one-fifth of the world's total. Australia and Israel are also serious producers. Israel has successfully addressed chronic water shortages through five large operational plants and has plans for five more. And in the United States, in drought-prone California, the largest desalination plant in the Western Hemisphere delivers 50 million gallons of fresh water to San Diego County from coastal Carlsbad. Next door in Orange County, a similar plant in Huntington Beach is expected to start delivering 50 million gallons a day by 2023. In 2022, California is home to 11 municipal seawater desalination plants, with 10 more proposed.

Investments in desalination plants will markedly increase as the availability of surface water and groundwater declines and the demand for fresh water increases.

### Continuing Development of Technologies

While the pressurized closed pipe flow has grabbed much of the attention, open channel flow measurement is an exciting area that is worth a closer look. Although it can be difficult to grasp at first because of the variety of methods, this variety also provides room for new technologies. As is the case with closed pipe flow, many of the recent developments revolve around electronic enhancements and improvements in communication. There is also room for truly novel developments involving new sensor technologies or the improvement of existing sensor technologies. Some improvements of this type have already occurred with Doppler technology. Many are watching for further developments as companies vie for pieces of a growing market.

## FACTORS LIMITING THE GROWTH OF THE OPEN CHANNEL FLOWMETER MARKET

While the future of open channel measurement appears brighter today than any time in the last several years, there are still factors that will limit the market's growth rate. Two of these are as follows:

- Delays in funding, planning, and deployment
- Lack of investment in open channel flowmeters

### Delays in Funding, Planning, and Deployment

While investment interest in open channel measurement has definitely intensified for the reasons stated above, the actual action steps needed to implement these best intentions may be some time in coming. Delays may occur despite the urgency of the need for several reasons.

The development of water and wastewater infrastructure development and/or improvements have traditionally been subject to lengthy funding and engineering cycles. These projects have frequently worked within 20- to 40-year engineering cycles. There are major cities in the United States, for example, that are using water distribution systems where portions were put into service more than 50 years ago. These old pipeline components are the source of outages and other inconveniences but have not been a high-priority item in most municipal or regional budgets. The practice of monitoring river systems and other surface water sources has also not been a high priority for most government budget makers, and the extension of private funding for this purpose has always been uncertain.

Unfortunately, this circumstance is likely to persist. In developing countries, the need to develop or upgrade freshwater supplies and adequately treat wastewater is subject to financial constraints and competing priorities. In developed countries such as the United States, the national backlog of infrastructure improvements exceeds the resources, including funding, materials, and people, to address all the issues immediately. The availability of these resources is the primary constraint on the further development of desalination plants.

### Lack of Investment in Open Channel Flowmeters

Even though there has been much progress in the development of open channel flowmeters in the past 10 years, the largest flowmeter companies are more focused on research and development in closed pipe technologies than on open channel flowmeters. Siemens and Endress+Hauser have maintained an active open channel research program in addition to their substantial research in closed pipe flowmeters. But some other major companies either do not offer open channel flowmeters or make them a very small part of their portfolio.

This is not stated as a criticism, but simply as a statement of fact. There is more demand for Coriolis and ultrasonic flowmeters that measure oil and gas than for open channel flowmeters that measure water. Regardless of the profit margins, the cost for Coriolis and ultrasonic flowmeters is much higher than that of open channel meters.

This way of thinking about research and development is likely to change over time as water shortages increase and the need to measure water flow becomes more critical. This is a case where market demand and market forces are likely to give the larger flowmeter companies more incentive to invest in open channel flowmeters. And smaller companies such as Rittmeyer in Switzerland and Teledyne Isco in the United States are already devoting much of their resources to open channel flow measurement. While this will not reduce the need for investment in closed pipe flow methods, open channel flow presents a growth opportunity for any company that wants to take it on. And with growing populations and a greater need to expand the water and wastewater industry, the future for open channel flowmeters is bright.

## FRONTIERS OF RESEARCH

The following are the frontiers of research for open channel flow measurement.

## Measuring the Size of the Channel

In many ways, open channel is more of an application than a technology, although it is treated as a technology in this book. The reason is that open channel flow does not refer to a specific technique of measurement. Instead, it is a mixture of different measurement technologies and devices, including level measurement, ultrasonic flow, radar, hydraulic structures, and other technologies and devices. What makes something an open channel measurement is that it involves measuring the water that flows by gravity rather than in a pressurized pipe. Open channels are by definition not closed, although measurement of water in partially filled pipes counts as open channel measurement. Open channel flow occurs when the different technologies and structures required for open channel flow measurement are harmonized in such a way as to produce a measurement of flow in an open channel such as a river, stream, or wastewater channel.

One of the most common methods of open channel flow is called the area-velocity method. This method uses the flow equation:

$$Q = A \times v$$

where $Q$ is volumetric flow and $v$ is velocity. $A$ refers to the area of the channel, river, or stream whose flow is being measured. One of the challenges of finding accurate open channel flow is finding a way to accurately measure the area of a river or stream. In most cases, the area will be something of an estimate because the body of the river or stream is not likely to fit exactly in the standard geometric structures such as circles and squares. Once this is known, and flow velocity is known, then the quantity of flow can be determined. However, time needs to figure into the equation because even if you know that water is flowing at 240 gallons per minute, you won't know how much water has flowed unless you know how many minutes it has been flowing. In this example, if water flows at 240 gallons per minute for 10 minutes, 2,400 gallons will have flowed in that period of time.

## Integrating Different Technologies

Part of the challenge for any open channel flowmeter manufacturer is to find technologies that work well together. In the area-velocity method, often ultrasonic or radar technology is used to determine the level of the water or stream. There are at least four different types of Doppler systems used to measure open channel flow velocity. Doppler transducers send out a signal that is reflected off of particles in the flowstream. The reflected signal has a lower frequency than the sent signal. By analyzing the differences in the sent and reflected frequencies, the open channel flowmeter is able to determine flowrate. Doppler is primarily an ultrasonic technology, but radar and laser Doppler techniques are also used to determine flow velocity.

Magnetic flowmeters are also used in open channel measurement. These are typically in the form of probes that lie at the bottom of the channel. Transit time ultrasonic flowmeters are used, but they are mostly used in partially filled pipes because the transducer needs to be attached to some structure. Also, transit time ultrasonic meters do best in clean water. If a river or stream has dirt and particles mixed in, then Doppler ultrasonic is a better choice.

Still another method of measuring flow velocity is the current meter. General Oceanics makes a current meter that is designed to measure the distance traveled by a moving object such as a boat. It works somewhat like a turbine meter in that it has a rotor-like blade that turns with the flow of a river or stream. A current meter can measure velocity if it is manually timed and the distance traveled is divided by the time. So, if the current meter shows that a boat travels 1,000 feet in 10 minutes, its velocity is 100 feet per minute.

## Integrating Flowmeters with Control Valves

The Clean Water Act of 1972 contained many environmental regulations governing the use and treatment of water. This includes treatment of stormwater runoff that runs into rivers and streams. Stormwater runoff is also being regulated in Europe. The stormwater runoff is held in a buffer area to prevent too much of it from flowing into the river at one time. This might be compared to an oil storage tank where crude oil is stored until it can be shipped to a refinery or moved to another location. Once it is safe to release more stormwater from the buffer area into the river, a magnetic flowmeter together with a control valve do the job together. The control valve opens to allow the runoff to flow, and the magnetic flowmeter measures how much stormwater is released.

This is one example of a broader trend. Just as mass flow controllers incorporate a valve, so more flowmeters are being sold with control valves and other instrumentation. In addition to valves, flowmeters are sold with pressure and temperature sensors and transmitters. Coriolis and positive displacement flowmeters pumps and valves are used on the back of trucks to dispense fuel oil and other refined fuel products. Oval gear meters are sold on skids for the purpose of custody transfer of crude oil. The trend is for flowmeters to be part of a system rather than just being isolated instruments. And even if they are isolated instruments, the trend is towards interoperability and more communication among the different types of instruments. The broad use of communication protocols is facilitating this communication.

# 9 Variable Area Flowmeters

## OVERVIEW

While variable area (VA) flowmeters are limited in their functionality, they cost far less than most other types of flowmeters. When users are looking for a simple, low-cost solution, VA meters are a particularly good fit. While they can measure flowrate, they are well suited for applications where a flow/no-flow determination is desired. They are also very effective at measuring low flowrates, and as they do not require electric power, they can safely be used in flammable environments.

## HISTORY

The history of VA meters dates to 1908 when they were invented by German engineer Karl Kueppers in Aachen, Germany. At that time they were called "rotameters," named after the rotating float that was originally a component of these meters. Felix Meyer recognized the commercial potential of Kueppers' invention, and in 1909 founded "Deutsche Rotawerke" in Aachen. The product invented by Karl Kueppers was the first VA flowmeter with a rotating float. The German company Deutsche Rotawerke was the forerunner of the company that was later known as the Rota Company. Originally, Meyer called his products "rotamesser."

In 1995, Yokogawa purchased the Rota Company and named the resulting company Rota Yokogawa. Rota Yokogawa still manufactures its VA meters, which it also calls rotameters, in Wehr, Germany. In the meantime, the GEC Crawley Company in Crawley, United Kingdom, began manufacturing the first glass VA meters and registered the name Rotameter as a trademark in the United Kingdom.

This name still exists as a trademark in the United Kingdom but has been passed down through a number of companies, including KDG Instruments and Solartron Mobrey. In March 2009, Emerson Process Management acquired Solartron Mobrey, presumably for its level, density, and flow computer products. As part of the acquisition, Emerson Process acquired the trademark to the name "rotameter" in the United Kingdom (Figure 9.1).

The terms "rotameter" and "variable area meter" have become synonyms over the years. It appears, though, that Yokogawa has the rights to the name rotameter in Germany while Emerson Process has the rights to the name in the United Kingdom. Regardless, the name rotameter has become a generic term for VA flowmeters, so its status as a true trademark is open to question. This is similar to what happened to "Xerox," which is now often used generically to describe any photocopy.

DOI: 10.1201/9781003130024-9

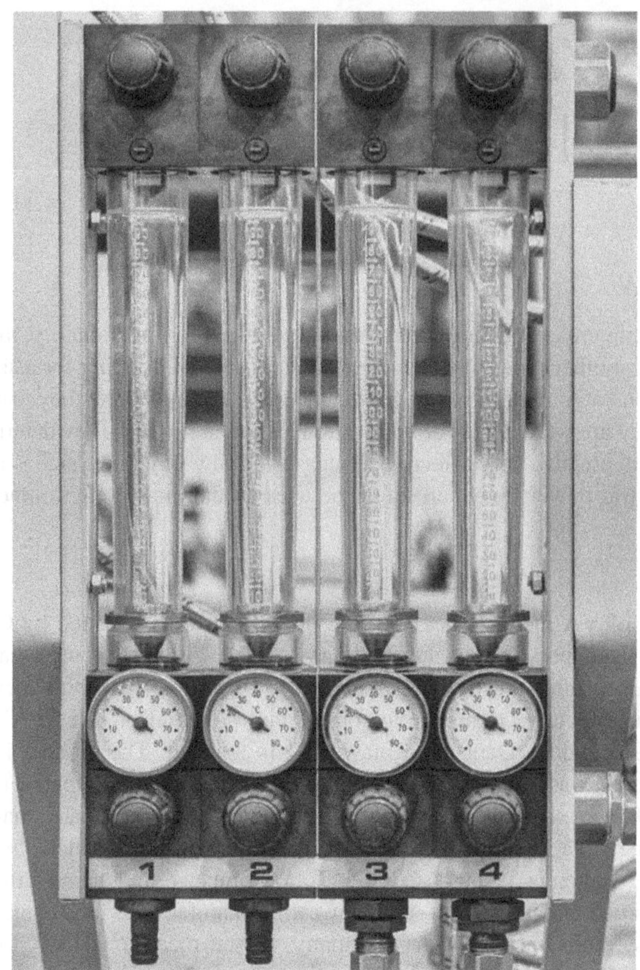


**FIGURE 9.1** A group of clear tube variable area flowmeters.

## VA FLOWMETER COMPANIES

### ABB

ABB is a global technology leader with a history spanning more than 130 years. In accordance with the company's evolving Next Level strategy, begun in 2014, ABB now maintains four business areas: Electrification, Robotics & Discrete Automation, Motion, and Process Automation (formerly Industrial Automation). ABB intends to be a significant driver of digital transformation within the industrial segments it operates, and each of its business areas is supported by ABB's innovative Ability™ digital platform.

In flow measurement, ABB manufactures and supplies magnetic, Coriolis, vortex/swirl, thermal mass, differential pressure (DP), and VA flowmeters, and primary elements associated with its DP offerings.

### History and Organization

ABB traces its roots to 1883 when the firm ASEA, a Swedish electrical industry company, was originally incorporated by Ludvig Fredholm. In 1891, a Swiss company, BBC (Brown, Boveri, and Cie), was formed by Charles Brown and Walter Boveri. BBC produced AC and DC motors, generators, steam turbines, and transformers. Over the ensuing decades, BBC and ASEA achieved many technological breakthroughs, which eventually led to the 1988 merger of the two companies and the creation of ABB as we know it today.

In January 2017, ABB announced its four business areas. The Process Automation business area (renamed from Industrial Automation beginning January 1, 2021) has five divisions: Energy Industries, Process Industries, Marine & Ports, Turbocharging, and Measurement & Analytics. Process Automation employed approximately 22,000 people and generated revenues of US$5.8 billion in 2020. The Measurement & Analytics division includes field instrumentation, flow, analytical, and force. Its 16 factories worldwide are supported by 34 service support centers. Altogether, the unit employs roughly 4,000 people, including over 600 dedicated service personnel.

### VA Flowmeter Products

ABB offers VA flowmeters in two product groups: the FAM (metal) series and the FGM (glass tube) series.

The FAM3200 is an armored purgemeter designed to meter small gas and liquid flows in chemical and pharmaceutical industries, gas analyzers, process systems, and well systems. It accommodates applications with cloudy, opaque, or aggressive fluids.

The VA Master FAM540 is a metal cone flowmeter for oil rigs and chemical plants that is also used in the pharmaceutical and food & beverage industries. It measures liquids, steam, and gases, and is ideal for aggressive and opaque fluids.

The FGM1190 glass tube VA meter offers a high degree of reproducibility – three ribs parallel to the center axis of the tube guide the float to make sure it stays centered in the tube. ABB offers a wide variety of float weights and meter tubes to exactly match the needed flow range.

The low-capacity glass Purgemaster FGM6100 for liquid and gas features a corrosion-resistant, high-strength stainless steel body; quick and easy snap-in tube construction; and an operator protection shield. A variety of materials and scale lengths optimize the meter's flexibility.

## Brooks Instrument

Brooks Instrument, a division of ITW and part of the parent company's Test & Measurement and Electronics business segment, manufactures a broad line of flow measurement and process control equipment in six product categories. The company offers plastic, glass, and metal VA flowmeters (rotameters). Brooks also offers Coriolis and thermal mass flow controllers and meters. In fact, it claims to have the world's largest installed base of mass flow controllers and to be one of the most trusted suppliers of mass flow technology to the global semiconductor industry.

Brooks' pressure and vacuum products include gauges, regulators, switches, transducers & transmitters, controllers, and direct vacuum measurement capacitance manometers. The company also offers vaporization products, semiconductor products, various accessories, and software. Brooks specializes in low-flow (2 inch / 50 mm or smaller) devices.

## History and Organization

Brooks Instrument, originally known as the Brooks Rotameter Company, was founded by Stephen A. Brooks in 1946 in Lansdale, Pennsylvania. Brooks later became the first instrumentation company in the world to be certified to ISO 9001 quality standards.

Brooks Instrument, acquired in 1964, became a part of Emerson's group of flowmeter companies in April 2001. This group included Rosemount, Daniel Industries, and Micro Motion. This affiliation remained until December 31, 2007, when Brooks Instruments was sold to American Industrial Partners Fund IV for approximately US$100 million.

In May 2009, Brooks acquired Key Instruments, Inc., a VA flowmeter manufacturer based in Trevose, Pennsylvania, offering precision machined acrylic and molded plastic, glass tube meters, and electronic flowmeters as well as flow control valves for medical, industrial, water, chemical, and laboratory applications. The Key Instruments product line was absorbed within the Brooks protocol of named/numbered instruments.

In January 2012, Brooks experienced its most recent ownership change when American Industrial Partners sold the company to Illinois Tool Works (ITW). The terms of the deal were not disclosed. ITW is a diversified industrial manufacturer of value-added consumables and specialty equipment with related service businesses. The company focuses on profitable growth and strong returns across worldwide platforms and businesses. Brooks is positioned within ITW's Test & Measurement and Electronics operating segment, which is one of the largest among ITW's seven segments.

## VA Flowmeter Products

Brooks is a leader in the VA flowmeter market. As one of the pioneering manufacturers in VA rotameter technology, Brooks' diverse portfolio of gas flowmeters is in use across virtually every industry. Brooks offers high-performance armored metal, glass tube, gas, and water flowmeters – from inline to low volume – for a variety of applications: basic liquid or gas flow measurement, rotating equipment flow measurement, process analyzers, high-pressure flow on offshore oil platforms, chemical injection, and purge liquid or gas metering.

The meters feature reliable, easy-to-read displays, fail-safe inline flow indication under any circumstance, an integral flow controller to compensate for varying pressures, materials and designs to suit multiple pressure ranges and chemical compounds, field-replaceable components and custom scales, and optional integrated flow switches, automated valves or alarms.

Brooks divested its line of plastic VA flowmeters to Zober Industries in July 2022. The glass designs provide increased accuracy. Metal meters can also accommodate steam flows, high/low pressure and temperature applications, and hazardous, remote areas.

## HOW THEY WORK

Most VA flowmeters consist of an upright, tapered measuring tube that contains a float. Fluid flows through the tube from bottom to top. The upward force of the fluid raises the float in the tube. The float is counterbalanced by the force of gravity. The point at which the float stays constant indicates the volumetric flowrate, which can be often read on a scale on the meter tube. VA meters vary according to their float shape and the material the float is made from. VA meters come in plastic, glass, and metal (Figure 9.2).

Their cost varies with the material of construction. Metal-tube meters are generally the most expensive and are designed for high-pressure and high-temperature applications. Plastic meters are typically the least expensive, followed by glass flowmeters. Some metal-tube meters can exceed $1,000 in price.

A group of VA meters called purgemeters have been designed for low-flow applications. For most plastic and glass VA meters, the flowrate can be read from the scale on the tapered meter tube. For metal VA meters, the position of the float is

**FIGURE 9.2** Variable area meters in an agricultural application.

magnetically transmitted to an analog indicator. These meters also have a scale indicating flowrate, and the position of the analog indicator displays the flowrate. These VA meters still rely on a float to indicate flowrate, and they also need to be read manually, like most of the tube-design meters. Metal VA meters can be used for liquid, gas, and steam applications.

## GROWTH FACTORS FOR THE VA FLOWMETER MARKET

Growth factors for VA flowmeters include the following:

- Users have a need for the measurements done well by VA flowmeters.
- The drive to enhance efficiency by measuring everything.
- Networking through communication interfaces.
- VA flowmeters are chosen by laboratory, research, and OEM users.
- VA flowmeters are a low-cost solution.
- Fast, regional delivery.

### USERS HAVE A NEED FOR THE MEASUREMENTS DONE WELL BY VA FLOWMETERS

While other flowmeters employ more complex principles, VA meters use a simple flow measurement principle. Many end-users still have a need for the type of measurement done by VA meters:

- Measure air and gases at low flowrates – a class of VA meters called purgemeters has been developed to handle a variety of low flow applications, including fluid that flushed out process piping
- Determine whether there is flow or no flow
- Activate an alarm if the flow exceeds a high or low limit
- Measure flow in high-temperature and high-pressure applications (metal and glass tubes)
- Easily check the performance of other meters
- Measure air, steam, compressed gas, and water
- Function without electric power; safe for flammable environments

VA flowmeters will continue to be sold because they fulfill important needs in the end-user community that are not going away, including the ones listed above. While this market is the slowest growing flowmeter type, it is still likely to show positive growth over the next 5 years.

### THE DRIVE TO ENHANCE EFFICIENCY BY MEASURING EVERYTHING

VA flowmeters are benefiting from the trend to optimize efficiency and recapture costs from users by measuring everything possible. They provide simple and inexpensive entry points for organizations and emerging countries that want to begin measuring flow but do not need high accuracy or large line sizes. (Most VA meters range from ≤ ½ inch to > 2 inches, with some as large as 4 inches.) In some situations, any measurement is better than no measurement. In other more established measurement applications, VA meters remain an inexpensive solution that is also easy to understand and maintain. Since VA meters are cheap and essentially disposable, they encourage people to measure at many points and even to consider previously unexplored applications.

VA flowmeters are used in the process industries, in research and laboratory environments, and in OEM applications. They are ideal for HVAC, industrial, and utility air and water flow. Another strong industry is Pulp & Paper, where organizations need a general idea of the flow or want to understand usage, but do not need precise measurements for billing or critical applications. In fracking, VA meters can indicate how much water an operator is using and returning to the ground. VA meters also find use on oil and gas platforms as they do not necessarily need to be powered.

## Networking Through Communication Interfaces

Although the majority of VA meters still need to be read manually in the field, some can now transmit data over two-wire circuits to a controller or recorder. Many VA suppliers today offer models that include 4–20 mA output with signals for up-to-date communication protocols, including HART, Foundation Fieldbus™, and Profibus®. Some also generate an output signal that can be sent to a controller or recorder. The development of meters with a transmitter output is an important advance in VA flowmeters, turning the meter into more than a visual indicator.

The ability to network VA meters, coupled with developments in diagnostic functions in general, means users can integrate VA flowmeters into their plant asset management systems. These more intelligent VA meters are becoming more valuable members of a company's measurement family. Ironically, these advancements are edging smart VA meters into competition with new-technology magnetic and ultrasonic flowmeters that offer the advantages of wireless technology and at the same time are becoming increasingly smarter and affordable. However, the price range of VA meters, particularly meters with plastic tubes, remains significantly lower than the price range for new-tech meters. And with low-cost manual VA meters, particularly with inexpensive plastic tubes, there really is no competition.

## VA Flowmeters Are Chosen by Laboratory, Research, and OEM Users

While VA flowmeters are used in the process industries, they are also widely used in research and laboratory environments. They are used in these environments to measure the flow of air and gases at low flowrates, when a visual indication is sufficient, to check on the performance of other meters, and when a low-cost measurement is desired. VA meters, including plastic meters, will also continue to be used for OEM applications.

## VA Flowmeters Are a Low-Cost Solution

Low cost is probably the most important advantage of VA flowmeters. The average selling price of VA flowmeters is in the hundreds of dollars compared to most other

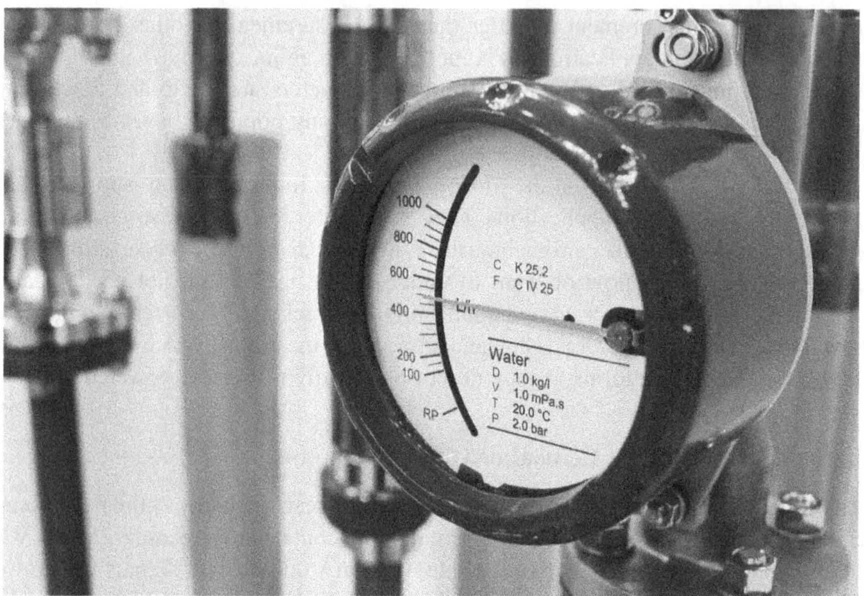


**FIGURE 9.3** A modular variable area (VA) meter: VA tube models of various materials are visible in the background.

types of flowmeters whose average cost ranges are in the thousands of dollars. While VA flowmeters are limited in their functionality, they cost far less than most other types of flowmeters. When users are looking for a low-cost solution, they will continue to consider VA meters.

The cost of VA meters varies with the materials of construction. Plastic meters are generally the cheapest, followed by glass flowmeters. Metal-tube meters are typically the most expensive and are used for high-temperature and high-pressure applications. Some metal-tube meters range in price between $1,000 and $2,000, though many are available for much less (Figure 9.3).

Users who simply want to determine a flow or no-flow situation, who need to set an alarm, or who want to check the performance of another flowmeter may select a VA meter to do the job. While many are read manually, some are now available with transmitters that have a 4–20 mA output. Expect end-users to continue to buy these flowmeters when they need a simple, low-cost flow measurement.

## Fast, Regional Delivery

In general, the VA market is a regionalized business. Users often order relatively large quantities of meters for prompt delivery and suppliers rely on local distributors to stock sufficient supply to meet their needs. As VA meters become more and more differentiated, suppliers are also turning to e-commerce sites to promote their business.

## FACTORS LIMITING THE GROWTH OF THE VA FLOWMETER MARKET

While VA flowmeters are an economical and versatile fluid measurement instrument, this device category has limitations that work against faster growth in the market and limit its use.

### Mounting Configuration

VA flowmeters can only be mounted in an in-line configuration. This limitation means that the commissioning of the meter will require a shutting down of an active process, a consideration when "uptime" is a deciding factor. Clamp-on ultrasonic flowmeters can be installed without process interruption, for example, and other hot-tapped insertion meter technologies also do not require process shutdown.

### Measuring Tube Material Choices

The end-user's choice of measuring tube is critical for success. High-temperature and high-pressure applications require metal tubes – and these are available in a variety of materials, including brass, aluminum, and stainless steel – but even these most robust of materials must be matched with the fluid properties to be measured.

When choosing a glass tube, it is important to know whether the meter will experience significant shaking (e.g., water hammer) that could damage the glass. Glass tubes are also susceptible to fluid properties such as high temperature or high pH levels that can soften some glass; or caustic soda which can slowly dissolve glass; or, certain liquids such as hydrofluoric acid that can etch glass, impairing its usefulness. And glass tubes are subject to breakage if connections are not made with special attention to the amount of torque applied when being fixed in place.

The point here is that particular attention must be practiced in tube material selection as this choice is dependent upon a number of factors beyond diameter and length in order to function satisfactorily over time.

### Fluid Pressure

VA flowmeters do not handle pulsating flows as well as alternatives and should be avoided altogether if the application will not always consist of a full pipe condition.

### Fluid Viscosity Changes

When measuring liquids, VA flowmeters do best with clean and consistent fluids. A prime illustration of this limitation is when a viscous liquid finds its way through the VA flowmeter leaving drag layers on the float. These layers will affect all subsequent measurements of the defined process fluid.

### Line Size

VA flowmeters are suitable for use in pipes with a maximum diameter of four inches. This limitation excludes their use in many water & wastewater applications, for example, where VA meters are otherwise a popular choice.

### Measurement Accuracy

Operators today are accustomed to higher standards being applied to every process measurement parameter. The accuracy of the flow measurement is one of the most basic and essential (in addition to the measurement of pressure, temperature, and level). VA flowmeters are typically used in applications where the highest accuracy (i.e., ≤1.0% of rate) is not required. Increasingly, this requirement is becoming more pervasive as the industry seeks higher process efficiencies, complies with stricter environmental regulations, and is challenged in other ways that demand the collection, analysis, and management of more precise data in real-time. VA flowmeters are severely challenged by this requirement whenever it arises.

### Applications

VA flowmeters are used in research and laboratory environments and in the process industries to measure the flow of gases and air at low flowrates. They are also used when a visual indication is sufficient, and to check on the performance of other meters. VA meters, including plastic meters, are also used for OEM applications. Plastic meters are applicable for water, gas, and air applications. The scales are separately marked for water and air applications, and they are available in both English and metric units. They are most widely used when the cost is the main consideration, and high accuracy is not required.

## FRONTIERS OF RESEARCH

The following are frontiers of research for VA flowmeters.

### Addition of Transmitters and Communication Protocols

One of the advantages of VA flowmeters is that they do not require power to operate. However, there is a broader trend within the flowmeter world for instruments to communicate with each other. Also, putting a transmitter with an output signal on a VA meter makes it possible to obtain a reading without manually reading the meter by comparing the float to the tick marks on the meter. This can make these meters available in situations where it is desirable to have the reading output to a recorder, programmable logic controller, or distributed control system but when high accuracy is not a requirement. There are also situations where VA meters serve like a switch, and simply register a flow/no-flow situation.

Adding a transmitter with an output to a VA meter does not prevent anyone from using these meters without power. Those types of VA meters are still available.

What it does is to add another option for end-users in case they want to use an upgraded VA meter.

## Putting the Right Scale on a VA Meter

Even though VA flowmeters are not sophisticated instruments, when compared to other types of flowmeters, they still need to be calibrated properly. Part of this issue has to do with the type of scale on the meter. Some scales are direct reading scales, and they have the units of measurement printed directly on the meter (e.g., Gal/min and L/min). The second type of scale has a series of tick marks on the meter but doesn't specify the units. This is called a correlated meter. Instead of printing the units on the meter, it has a series of tick marks, for example, 0–65 or 0–150. It comes supplied with correlation sheets that correlate the tick marks to a variety of engineering scales. One advantage of the correlated meters is that the same meter can be used to measure both liquids and gases. As liquids and gases are measured in different units, the same direct reading scales cannot be used for both of them. Another advantage of correlated meters is that if temperature and pressure conditions change for an application, the user can use a different correlation sheet to reflect the new parameters. If temperature and pressure conditions change for a direct reading meter, it may need to be sent back to the manufacturer for recalibration.

## Dealing with Calibration Issues for VA Flowmeters

Calibration is an issue for almost every type of flowmeter, and VA meters are no exception. Flowmeter accuracy can vary quickly if the temperature and pressure conditions vary from the standard calibration conditions. They vary more for VA meters designed to measure gas flow than they vary for those designed to measure water flow. This is because density and viscosity change very little for liquids with temperature and pressure changes. Typically, manufacturers calibrate their gas VA meters at a standard temperature and pressure, which is 70°, and with the meter open to ambient conditions.

It may be that many end-users do not care about recalibrating their VA meters, or they figure that the accuracy is low anyhow so it's not worth worrying about. However, there are many end-users who rely on their VA flowmeter readings, and who perhaps should be educated about a potential need for recalibration, or who should be provided with a means to recalibrate their VA meters.

Manufacturers of other types of flowmeters have provided ways to self-verify the correctness of a meter reading without sending the meter out for calibration. This is certainly true for manufacturers of Coriolis meters. Certainly, someone who has a correlated VA meter will have an easier time adjusting to varying temperature and pressure conditions than someone with a direct reading meter. Perhaps, VA suppliers can come out with "calibration kits" to enable end-users to calibrate their VA meters without sending them out for recalibration. This is another frontier of research for VA flowmeters.

# 10 Flowmeters and the Oil and Gas Industry

## OVERVIEW

The oil and gas industry is one of the largest industries in which flowmeters are used. Both oil and natural gas need to be measured at many points within the process stream that begins with drilling and ends with the delivery of refined products to customers. Many other types of instrumentation are needed, including temperature and pressure transmitters and sensors, analytical equipment, level products, as well as measurement and control equipment.

It is important to understand the different types of fluid measured when looking at the different types of flowmeters. Ultrasonic flowmeters that measure petroleum liquids cannot also measure natural gas because the speed of sound is different for these different fluids. Some flowmeters, such as thermal meters, almost exclusively measure gas, whereas others, such as magnetic flowmeters, almost exclusively measure liquids. End-users will pay from $75,000 to $100,000 for flowmeters that measure high-value liquids such as crude oil and natural gas but will not pay that much for a flowmeter that measures water. Understanding the different types of fluids that flowmeters measure is fundamental to understanding the structure of the flowmeter market.

This section defines and explains the following types of fluids that flowmeters measure:

- Petroleum liquids
- Non-petroleum liquids
- Gases
- Industrial gases
- Natural gas

## TYPES OF GAS AND OTHER FLUIDS

### Petroleum Liquids

One important distinction within liquids is between petroleum and non-petroleum liquids. Petroleum liquids are a broadly defined class of hydrocarbon mixtures. A hydrocarbon liquid contains both hydrogen and carbon atoms. Two main classes of petroleum liquids are crude oil and refined liquids. Crude oil comes out of the ground as a result of the drilling process. When it comes out of the ground, it generally is mixed with gas and water. These three fluid components are physically separated in what is called test and production separators. In this upstream part of the process, many differential pressure, ultrasonic, and turbine meters are used. Eventually, the

DOI: 10.1201/9781003130024-10

crude oil is sent to a refinery, where it is separated into its component parts. The result of this distillation and refining process is refined liquids, including refined fuels.

The distinction between petroleum and non-petroleum liquids is important in flow measurement because petroleum is a high-value commodity and often high accuracy is required when measuring it. Viscosity is an important property in measuring petroleum liquids, along with flowrate, temperature, and pressure. There are many places downstream of the refinery where refined liquids are measured. Eventually, these liquids reach the consumer in the form of gasoline, diesel fuel, motor oil, ethanol, methanol, and many other types. Some flowmeters are specifically designed to measure petroleum liquids. Magnetic flowmeters are unable to measure the flow of any petroleum liquids.

## Non-petroleum Liquids

Non-petroleum liquids include water, juice, milk, honey, tomato sauce, wastewater, and many other types. Hygienic and sanitary conditions are especially important at food processing plants. They are also important at breweries and microbreweries, which use many magnetic and Coriolis flowmeters (Figure 10.1). Water and wastewater treatment plants use magnetic, differential pressure, and ultrasonic meters to measure both clean water coming into and wastewater going out of the plant. The process of delivering clean drinking water to millions of people requires many measurements from the source of the water through the clean water plant down to delivering water to the myriad homes, businesses, and industrial plants that populate a typical urban area.

**FIGURE 10.1** Tanks at a beverage processing plant.

## Gases

Even though we can't see them, gases are all around us. Gases are distinguished according to their different atomic configurations. While there are some "pure" gases, such as hydrogen and oxygen, there are also many gas mixtures. Some are simply combinations of different atoms into molecules such as carbon dioxide ($CO_2$) and water ($H_2O$). The air we breathe is a mixture of gases, primarily nitrogen, oxygen, and argon along with some other trace gases. Some mixtures of oxygen, nitrogen, and carbon dioxide are used to delay food spoilage and increase its shelf life.

There are many types of gases. Some types of gases, apart from natural gas, include the following:

- Industrial gases
- Fuel gases
- Medical gases
- Specialty gases
- Welding gases
- Breathing gases
- Syngas
- Landfill gas
- Biogas
- Shale gas

## Industrial Gases

Industrial gases are specific types of gases that are gaseous at ambient pressure and temperature and are specifically manufactured for use in industry. Any gas that is put in a canister is considered an industrial gas, with the exception of certain fuel gases. In fact, industrial gases are typically put into canisters or cylinders and sold for use in that form. Some of the better-known companies that manufacture industrial gases include Air Liquide, Air Products & Chemicals, The Linde Group, Nippon Gases, and Praxair.

The mere fact that a gas is used in industry does not make it an industrial gas. Some of the other types of gases listed above are used in industry, but they are not considered to be industrial gases. This is mainly because they are not manufactured and packaged in canisters or cylinders to be used for specific purposes. This excludes medical gases and breathing gases, but most of the remaining gases would have some use in the process industries.

## Natural Gas

The US Energy Information Administration (EIA) defines natural gas as "A gaseous mixture of hydrocarbon compounds, the primary one being methane." Natural gas is called "natural" because it occurs in nature. In fact, its occurrence in nature is always underground, whether it is under land or under water. Some natural gas occurs deep below the ocean depths, and this has led to the development of some very advanced drilling involving subsea technology.

Like oil, natural gas developed over millions of years as a result of the decay of millions of microorganisms. Both natural gas and oil were created over vast amounts of time as a result of the combination of temperature and pressures on these microorganisms. It is unfortunate that we are using up these vast resources that took millions of years to develop in the course of several 100 years. Fortunately, many companies have turned to renewable fuels as an alternative source of energy. These may not replace oil and gas, but they will allow them to survive for use by future generations for applications that cannot be satisfied by renewable energy.

The hydrocarbon fluid that comes out of a wellhead is a good example of a fluid mixture, but it is not typically only a gaseous mixture. In most cases, what comes out of a wellhead is a mixture of crude oil, natural gas, and water. These components are separated by test and production separators and sent off to different locations. Typically, the natural gas goes to a gas processing plant, while crude oil goes either to a refinery or into storage. At the gas processing plant, the natural gas is stripped of any natural gas liquids (NGL) that may still be associated with the natural gas. At this point, the natural gas is almost entirely pure methane, and it is called "dry" natural gas.

## THE THREE MAIN SEGMENTS OF THE OIL AND GAS INDUSTRY

The oil and gas industry is often divided into the following segments:

- Upstream
- Midstream
- Downstream

What they have in common is that all involve different processes that oil and gas go through as they pass through each of these segments. Flow measurement plays a vital role in all three of these segments.

### UPSTREAM

Oil and natural gas are found and located in the upstream segment. Geologists use many different methods to determine the location of oil and natural gas reservoirs. These include seismic surveys and testing, collecting and analyzing rock samples, and the use of satellite imagery.

Once petroleum reserves are located, drilling rigs are used to drill oil wells to find the oil reservoir. A drilling rig contains a rotating shaft that is used to drill a borehole down to the desired depth. As the drilling occurs, steel pipe called casing, held in place with cement, is used to prevent the sides of the well from caving in. Due in part to the high underground pressures, drilling a well requires serious attention to safety issues.

Some wells are more productive than other wells. Over one million oil and gas wells have been drilled in the world, and over half of them are in the United States. But many of these are low-producing land-based wells. Subsea wells like those located in the Gulf of Mexico typically have higher flowrates but are more expensive to drill. As a result, they produce more oil than land-based wells. The Middle East has fewer wells than the United States, but the wells there are much more productive. For example, Saudi Arabia's Ghawar oil field is famous for producing five million barrels of petroleum per day for decades. This is a record that may never be surpassed.

Once an oil well is drilled, and production begins, underground pressures push the fluid containing oil to the surface. This fluid is typically a mixture of oil, gas, and water. Once it reaches the surface, it passes through test separators and production separators. These separators physically separate the oil, gas, and water, and send them to different locations.

The upstream oil field has many flow measurement opportunities. In some cases, multiple wells feed into a common flowstream, and the fluid amount has to be measured from each well. This is sometimes done by a process called allocation metering. Fluid is measured as it enters and leaves the test and production separators. Often ultrasonic, differential pressure, or turbine meters make these measurements. Coriolis meters do better with liquids than with gas and they also have limitations in terms of line size.

## MIDSTREAM

The midstream portion of the oil and gas process stream involves transporting the oil or gas from the upstream oil field down to a gas processing plant or a refinery. This can be done by truck, railcar, ship, or pipeline. Oil storage is also part of the midstream segment. Often crude oil from an upstream oil field is not yet ready to be processed, so it is stored in large oil tanks that serve as a holding area until it is needed. One of the most famous crude oil storage areas is in Cushing, Oklahoma, which stores about 90 million barrels of oil.

Where they are available, oil and natural gas pipelines are often the preferred methods of transportation. TransCanada Corporation, which is headquartered in Calgary, Canada, transports natural gas through a vast network of pipelines to destinations in Canada and the United States. Many states in the Northeast are included in these destinations, though the pipelines extend as far south as the Gulf of Mexico.

Custody transfer of natural gas, especially for large natural gas pipelines, is one of the fastest growing applications in the flowmeter market. Ultrasonic, differential pressure, and turbine are the main types of flowmeters used for this. Ultrasonic flowmeters have been gaining ground because they do not have moving parts and are non-intrusive. They are also highly accurate, and those with three or more paths typically meet industry guidelines for accuracy.

Crude oil and also refined fuels often travel by train or by truck to areas where pipelines do not exist. Trains offer high-speed delivery of crude oil throughout the United States, although like pipelines, there are many places they don't go.

New areas such as North Dakota's Bakken region have begun producing oil with the advent of shale drilling and fracking. These are regions where pipelines may not already exist, or where putting in additional pipelines is not feasible. Texas and North Dakota are the leading states that use railcars to transport crude oil.

Trucks are often used because they can go almost anywhere in the United States, given its vast network of highways. Trucks play a major role in delivering refined fuels to their point of use. Trucks are still the typical means for delivering fuel oil to businesses and homes that use oil as a source of heat, since it is often the most practical delivery method. Either positive displacement or Coriolis meters are often used on the back of these delivery trucks. These flowmeters are part of an integrated system that includes pumps and valves.

Besides trains, trucks, and pipelines, ships are increasingly important in the delivery of crude oil and refined products. Oil tankers carry crude oil to places that do not have pipelines. Oil tankers often carry crude oil to refineries. The Strait of Hormuz serves as a gateway for many ships from the Middle East on their way to the Japan, China, Western Europe, and the United States. The Panama Canal is another critical passageway for ships carrying crude oil from the United States to Latin America and to other locations in the United States.

## Downstream

The downstream part of the oil and gas process stream is downstream from a gas processing plant or a refinery. Crude oil is transported to a refinery via railcar, truck, ship, or pipeline. The refinery converts the crude oil into different refined fuels, such as diesel, kerosene, gasoline, jet fuel, and fuel oil. These refined fuels are often transported to their next destination via pipeline. Trucks typically deliver the products to their point of use, especially fuel oil and gasoline.

Refineries offer many flow measurement opportunities. As the crude oil goes through the distillation process and is converted into various types of refined fuels, there are many points of measurement. Much of this does not require custody transfer accuracy since it is intra-plant measurement. Within refineries, differential pressure, ultrasonic, turbine, and vortex meters are all used for flow measurement. In case steam measurement is required, differential pressure and vortex meters are favored. The crude oil entering the refinery is measured, and the refined fuels that leave the refinery are also measured. Typically, these are custody transfer measurements.

Natural gas processing plants work in an analogous way to refineries. They use natural gas as feedstock instead of crude oil. A gas processing plant strips off the impurities and non-methane hydrocarbons to produce natural gas that is considered "pipeline quality." As part of this process, natural gas liquids (NGLs) such as ethane, propane, and butane are recovered. These NGLs, which have multiple uses including enhanced well recovery, are sold separately. Ethane is a feedstock for ethylene, which has a wide variety of uses.

Industrial gases are made up of compounds, elements, or mixtures. Nitrogen, hydrogen, oxygen, and carbon dioxide are common examples. A limited number of companies manufacture them, including Air Products and Chemicals, Air Liquide, and Praxair. Many industrial gases are delivered in cylinders to the retail and healthcare markets. Some industrial gases are delivered by pipeline or by truck. They are also used in chemical, pharmaceutical, and food & beverage plants.

Flow, temperature, and pressure measurement opportunities abound in natural gas processing plants. As natural gas enters the plant, it is measured. This is typically a custody transfer measurement. Many in-plant measurements do not require custody transfer accuracy, as is the case with refineries. Flowmeters that do well with gas flow such as vortex, differential pressure, thermal, and turbine meters are most likely to be used for in-plant flow measurement. Plants that manufacture industrial gases offer similar opportunities.

## THE OIL AND GAS INDUSTRY AND THE PRICE OF OIL

While flow measurement opportunities abound in the oil and gas industry, they are not always equally available. Much depends on the price of oil. For oil producers to be able to make a profit in producing oil, they need to sell it for an amount that exceeds their cost of production. The price of oil has a ripple effect throughout the oil and gas industry. When oil prices are high, oil production is profitable and companies can afford the cost of exploration and production, as well as the more expensive subsea drilling. All of this activity is instrumentation intensive. When prices are low, oil producers have to curtail their exploration and production activities. If less oil is being produced, there is less available to be refined and to be distributed downstream.

Because the demand for flow measurement is so closely tied to oil prices, the rest of this chapter looks at the price of oil from 2014 until the present. The downturn in the oil and gas industry in 2016 had a significant impact on the demand for flowmeters during this time. Likewise, the COVID-19 pandemic in 2020 had a major effect on the need for flowmeters at the upstream, midstream, and downstream levels.

Understanding this story requires looking at the Organization for Petroleum Exporting Companies (OPEC), its attempts to stabilize prices, and also the role of the United States and other major producers including Russia. The story is a fascinating one, and one that has an impact today, even as the world copes with new instabilities in Europe and in the supply chain. Despite all these issues, the world's need for energy will not be diminished, and it will have to be satisfied one way or another. As some of Russia's oil is taken off the market, and Europe looks for other sources of energy, liquefied natural gas (LNG) can be expected to become increasingly important as an energy source. The United States, Australia, and other countries can be expected to step in to fill the void created by the reduced supply of oil. This is an ongoing story that requires close watching.

## THE UNDERLYING DYNAMICS OF THE OIL MARKET

This section explains some of the structures that play a role in understanding the dynamics of oil prices. It begins by talking about the role of hydraulic fracturing, or "fracking," and the effect it has had on oil prices. The advent of fracking has increased the supply of oil, potentially having a depressing effect on oil prices. It then explains that there is no such thing as "the price of oil," since there are more than 150 different types of oil in the world. This section focuses on four types that have become "benchmarks" for oil prices in different geographic regions. It then discusses the formation of OPEC, why it was formed, and OPEC's role in controlling oil prices since 1970.

The price of crude oil in barrels has been of major concern to the flowmeter and other instrumentation suppliers for many years. The oil and gas industry is a major consumer of flowmeters, temperature sensors, pressure transmitters, and other instrumentation products at many different phases along the process stream from wellhead to distribution points.

From 2011 to August 2014, oil prices for the most part remained in the range of $80 to $100 per barrel. During this time, the world demand for oil exceeded the world's supply, and the support was there for relatively high oil prices. This spurred exploration and production worldwide, benefiting suppliers of flowmeters and other instrumentation.

Beginning in August 2014, worldwide supply began to exceed worldwide demand. This began a downward spiral in oil prices that resulted in oil prices bottoming out at just above $26 per barrel in February 2016.

## The Role of Fracking

Why did the world oil supply increase in 2014 and 2015, pushing down oil prices? Many analysts point to the advent of hydraulic fracturing, or "fracking," which makes it possible to get more oil out of difficult formations or existing wells than was previously possible. Fracking, which is often used with horizontal drilling (but is a distinct method), helps bring additional oil and gas to the top of the well. Until recently, fracking was largely an American phenomenon, but this technology is spreading to other countries. Some other countries where fracking is practiced include Canada, the United Kingdom, Poland, China, and New Zealand.

What is fracking? Hydraulic fracturing involves forcing a liquid (mainly water, usually containing sand and additives) through a well and against a rock formation until it fractures. The liquid is under high pressure. As the high-pressure liquid in the wellbore flows into the formation, the fracture extends deeper into the rock. When the injection is stopped and the pressure is reduced, the formation attempts to return to its original configuration. However, the formation remains open due to the sand and chemical-containing liquid in the well. As a result, hydrocarbons such as crude oil and natural gas flow from the rock formation into the well and can then be brought up to the surface.

After bottoming out in February 2016, oil prices retraced their steps through the $30 per barrel to $40 per barrel range. During the first three quarters of 2017, the price of crude oil ranged between $45 and $55 per barrel. Then in September 2017, prices remained firmly above $50 per barrel, settling above $60 per barrel on December 29, 2017. Oil prices remained above $60 for most of 2018, and at times sold above $70 per barrel. It wasn't until November 2018 that prices fell back below $60 per barrel.

## Types of Oil and the Formation of OPEC

It is possible to read the changes in oil prices from a table and not look further into the market dynamics behind those changes. However, if we are to understand the future of oil prices, it is highly instructive to look at the past. This section explains how oil came to be measured in 42-gallon barrels. While there are more than 150 types of crude oil in the world, the next section focuses on several that are used as "benchmarks," including West Texas Intermediate (WTI) and Brent crude oil.

Oil prices are very much a result of the balance between supply and demand on a worldwide basis. While changes can occur due to specific regional events, oil prices as a whole are a global phenomenon. What is somewhat regional is the differences among the different oils that are used as benchmarks for the price of oil. One benchmark oil is WTI, based in Texas and surrounding states. WTI is the benchmark oil for the United States. The price of Brent crude oil is a European and international standard for the price of oil. Brent is based in the North Sea, between Norway and the United Kingdom.

## How Oil Is Measured

Oil is measured in 42-gallon barrels. The history of this tradition goes back to 1866, soon after Edwin Drake drilled the first oil well in the United States in Titusville, Pennsylvania, in 1859. In 1866, a group of independent oil producers met in Titusville and decided that the 42-gallon barrel was the best way to transport oil. At that time, barges floated barrels of oil down the Allegheny River to Pittsburgh on the way to be refined into kerosene. The adoption of this standard for oil measurement stuck, and today oil is still measured in 42-gallon barrels (Figure 10.2).

**FIGURE 10.2** Modern oil "barrels".

Oil production is typically measured according to how many 42-gallon barrels are produced in a day. In many cases, this amounts to thousands or millions of 42-gallon barrels per day.

## Factors that Influence Oil Prices

The most fundamental determinant of oil prices is supply and demand. When the demand for oil exceeds supply, oil prices tend to rise, or to remain high, on a relative basis. When supply exceeds demand, however, oil prices tend to decline, or remain low, on a relative basis.

Of course, there are many other factors that influence the price of oil in addition to the balance of supply and demand. These include currency fluctuations, sudden disruptions in major sources of supply, political factors, bad weather such as hurricanes, disasters such as oil spills, etc. All these factors can cause oil prices to spike, or plummet on a temporary basis. Usually, though, these effects are temporary and oil returns to the price dictated by the balance of supply and demand. Oil prices also depend on the type of oil.

## Four Benchmark Oils: WTI, Brent, Dubai/Oman, and the OPEC Reference Basket

Before exploring the effects of supply and demand on today's oil markets, it is worth taking a look at what is meant by "the price of oil." While there are many types of oil, four types have become benchmarks for the oil markets. These are WTI, Brent, Dubai/Oman, and the OPEC Reference Basket.

**WTI** is traded on the New York Mercantile Exchange (NYMEX). It is composed of oil extracted in the United States, mainly from fields in Texas, North Dakota, and Louisiana. WTI is light and sweet and has a low sulfur content. WTI is extracted from oil fields in the United States and transported via pipeline to Cushing, Oklahoma, where it is refined. The price of WTI is a benchmark for oil sold in the United States.

**Brent** crude oil is extracted from oil fields in the North Sea. Originally, it was named after oil extracted from the Brent oil field, which is located off the coast of the United Kingdom in the North Sea. Today, Brent is mainly extracted from four oil fields in the North Sea: Brent blend, Forties blend, Osberg, and Ekofisk. While it is considered to be both light and sweet, it is slightly heavier than WTI. Brent futures are traded on the ICE Futures Europe in London. The price of Brent crude is a benchmark for oil produced in the North Sea and sold in Europe, Africa, Australia, and some Asian countries (Figure 10.3).

**FIGURE 10.3** A Brent Goose – Photo courtesy by Andreas Trepte, www.photo-natur.net.

Esso and Shell named the North Sea oil fields in the order of discovery after seabirds, alphabetically. Brent was the second oil field discovered and was named after the Brent Goose in 1972. The Brent Goose is a small goose with a short and stubby bill.

For many years, Brent and WTI traded at roughly the same amount. Then in 2011, when oil prices increased, Brent began trading higher than WTI. Although the reasons for this are debatable, some analysts attribute it to the fact that the North Sea oil fields are being depleted, while WTI in the United States is more plentiful. Canadian oil production is also increasing. Production from the Brent field has declined to the point that in early 2017, Shell announced plans to decommission this field over time.

**Dubai/Oman** oil refers to a "basket" of oils from Dubai, Oman, and Abu Dhabi. As a benchmark, it is an average of the prices of oil from Dubai, Oman, and Abu Dhabi. It is heavier than WTI and Brent oil and is slightly sour. Dubai/Oman oil has been traded on the Dubai Mercantile Exchange since 2007. It has become a benchmark for oil shipped to Asia.

The **OPEC Reference Basket** is another benchmark for oil prices. This is a blend of oils from most of the OPEC countries. The value of this Reference Basket is calculated by the OPEC secretariat in Vienna, Austria. It includes oil from Saudi Arabia, Iran, Qatar, Kuwait, and a number of other OPEC countries.

## What is OPEC?

To many people, OPEC is just part of the international framework on oil prices. Many people have a negative impression of the role that OPEC has played over the years in controlling oil prices. In reality, OPEC has played a very positive role in controlling oil prices (Figure 10.4). What is OPEC, and when was it formed?

OPEC is a permanent, intergovernmental organization composed of 14 major oil-producing countries. It was founded at the Baghdad Conference on September 10–14, 1960, by the following countries:

- Iran
- Iraq
- Kuwait
- Saudi Arabia
- Venezuela

The following ten countries joined in the intervening years. The year beside the country indicates the year they joined OPEC:

- Qatar (1961)
- Indonesia (1962–2016)
- Libya (1962)
- United Arab Emirates (1967)
- Algeria (1969)

**FIGURE 10.4** Flags of OPEC and three of the founding countries, Iran, Iraq, and Kuwait.

- Nigeria (1971)
- Ecuador (1973)
- Gabon (1975–1994, 2016)
- Angola (2007)
- Equatorial Guinea (2017)

Gabon was a member of OPEC from 1975 to 1994 but withdrew when it was unable to get a reduction in its annual fee. At that time, the annual fee for being a member of OPEC was $1.8 million. Since that time, it has increased to $3.1 million annually. Even so, Gabon did rejoin OPEC on July 1, 2016. Indonesia suspended its membership in 2009, also to save the annual fee, but rejoined the organization at the beginning of 2016. However, on November 30, 2016, Indonesia was suspended from membership in OPEC.

Initially, OPEC's headquarters were in Geneva, Switzerland. On September 1, 1965, the organization moved its headquarters to Vienna, Austria.

OPEC's stated objective is

> To co-ordinate and unify petroleum policies among Member Countries, in order to secure fair and stable prices for petroleum producers; an efficient, economic and regular supply of petroleum to consuming nations; and a fair return on capital to those investing in the industry.

OPEC's statute calls for it to have two ordinary meetings a year to decide any policy issues. However, OPEC meets in extraordinary sessions when required.

## Why Was OPEC Formed?

OPEC was formed in 1960 in response to import quotas on oil. In 1959, the US government established a Mandatory Oil Import Quota Program that restricted how much crude oil and refined products could be imported into the United States. This program gave more favorable terms to imports from Mexico and Canada. As a result, countries in the Persian Gulf received lower prices for their oil. Venezuela was another country that was negatively impacted by the Mandatory Import Quota Program.

In September 1960, Saudi Arabia, Iran, Iraq, Kuwait, and Venezuela met in response to the US Import Quota Program to form OPEC. Their goal was to obtain higher prices for crude oil. In 1960, crude oil was selling for $1.63 per barrel. This is a price that is difficult to imagine today when oil is selling in the $100 per barrel range. OPEC was largely unsuccessful in obtaining higher oil prices during the 1960s. However, in 1973, the organization was able to raise oil prices by curtailing production. Since then, this has become OPEC's main tool for influencing prices. OPEC controls a sufficient amount of oil production worldwide that it can raise prices by cutting production, thereby reducing the available oil supply.

## OPEC and "The Seven Sisters"

When OPEC was formed, the world was not exactly operating under the principles of free trade. Instead, a group of major oil companies that were informally called "The Seven Sisters" cooperated to control much of the world's oil production and distribution, along with oil prices.

According to various accounts, the origin of "The Seven Sisters" goes back to an agreement signed on September 17, 1920, among Royal Dutch Shell, Anglo-Iranian, and Standard Oil (now Exxon). Its primary purpose was to control oil prices.

In the following decades, other companies joined this group, and by the 1950s, this group was composed of the following seven companies (Figure 10.5):

- Exxon
- Mobil
- Chevron
- Texaco
- Gulf Oil
- Shell
- British Petroleum

These are all familiar names. This group controlled the distribution of crude exports throughout the world through its ownership of many of the major pipelines in the world. Many members of this group were also partial owners of

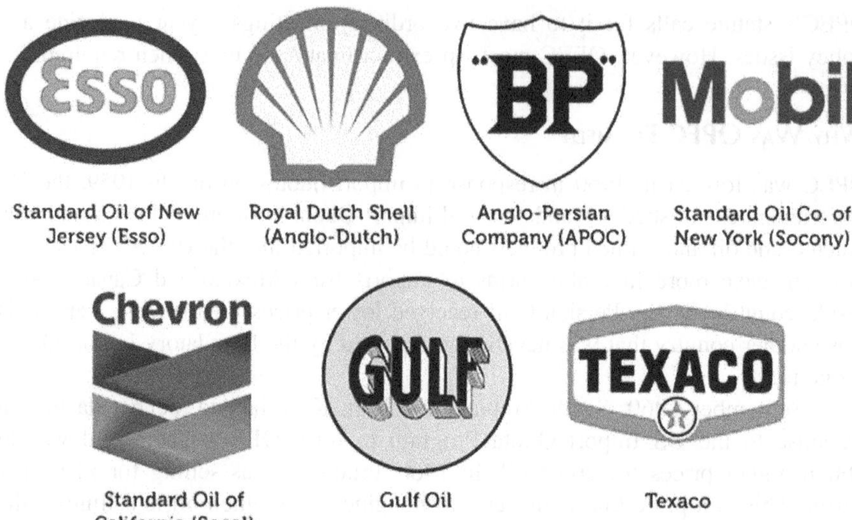


**FIGURE 10.5** Logos of the "Seven Sisters" companies.

the major oil companies in the Middle East, such as Saudi Aramco and Kuwait Oil Company.

OPEC is correctly called a cartel, but, in reality, when OPEC was formed, a major portion of the oil in the world was under the control of another cartel called "The Seven Sisters." Hence OPEC, in its formation, was simply forming an organization somewhat similar to one that already existed and that had operated for many years.

## OPEC Since 1970

Oil prices have been volatile on a number of occasions since their spike in 1970. In 1973, oil prices spiked due to the Arab oil embargo, and again in 1979 because of the Iranian Revolution. Prices were high in the early 1980s but crashed in 1986 due to an oil glut and a lack of consumer demand. Later in the 1980s, OPEC was instrumental in bringing prices back up by introducing a group production ceiling. Even so, oil prices were only about half as much as they were in the early 1980s. In 1980, the average price of a barrel of oil was in the $35 range; in 1989, it was $17 per barrel.

In the 1990s, oil prices remained relatively stable: between $15 and $22 per barrel. In 2004, oil prices began rising again. This continued until July 2008, when they peaked at $147 per barrel. Then an economic collapse and recession cut prices by more than half in 2009. However, by 2011, oil prices had recovered to the $100 per barrel range. Oil prices remained in the range of $80–$100 per barrel until August 2014. This was the month when worldwide oil supply began to exceed worldwide demand, and prices began to decline.

## HOW SUPPLY AND DEMAND IMPACT OIL PRICES

When oil prices are compared to the balance of supply and demand, it seems pretty clear that the decrease in oil prices correlates quite strongly with the imbalance in the supply/demand equation. On the demand side, economic weakness tends to generate weakness in demand, which can quickly lead to an increase in supply. This is likely to drive oil prices down. Many analysts point to reduced demand from the Chinese economy as a major factor in reducing demand.

There are other factors on the demand side. Automobiles are becoming more efficient, requiring less gasoline, and many companies are shifting to natural gas as a cleaner alternative to oil. Despite recovering economies, factors that influence reduced demand are still at work. While the implementation of clean and renewable energy is still in its early stages, it is already having an impact on the amount of oil needed by many economies.

However, many analysts point to the supply side as the main reason for the imbalance in supply and demand. The advent of hydraulic fracturing, or "fracking," has greatly increased the crude oil output of a number of countries, especially the United States. Hydraulic fracturing has made it possible to get more oil out of existing wells and to obtain oil from wells that were once thought to be "dry" or no longer viable.

From April 9-12, 2020, OPEC met to deal with declining oil prices. This action was made necessary due to a disagreement between Saudi Arabia and Russia about oil production amounts that goes back to a March 9, 2020 Extraordinary OPEC meeting. The timing of this meeting unfortunately coincided with the time when the effects of the COVID-19 pandemic began to seriously curtail demand for refined petroleum products. COVID-19 also depressed demand in January and February 2020, but it was in March that these effects became very serious as countries went into a state of lockdown and many business activities were curtailed.

Because it costs many producers $35 to $40 to produce one barrel of oil, prices below $30 mean that these producers cannot operate at a profit. If prices are sustained at this low level, some producers will be forced to go out of business. While the historic April 12 agreement reached at the OPEC meeting did not guarantee higher prices, it at least adjusted oil supplies to compensate for the plummeting demand brought on by COVID-19. When people are forced to stay at home, they don't drive to work, don't go out to eat at restaurants, and have to curtail vacation plans. This sudden drop in demand is what caused increased instability in oil prices.

## THE FLOWMETER MARKET IN 2020

The flowmeter market in the first half of 2020 was affected by two major negative factors:

- A decline in oil prices
- The effects of the COVID-19 pandemic

The flowmeter market was up modestly in 2019, after having a very strong year in 2018. However, the combined effect of the above two factors had a negative impact on the oil and gas industry, along with many other industries that are regular consumers of flowmeters. These include chemical, food & beverage, pharmaceutical, power, water & wastewater, and other process industries. As a result, the flowmeter market declined in 2020.

## FLOW MEASUREMENT AND OIL PRICES

There are many opportunities for flow measurement in upstream, midstream, and downstream oil and gas. This helps explain why the flow measurement market declined in 2015 after oil prices went from over $100 per barrel in 2014 to under $30 per barrel in 2016. Many oil exploration and production projects were canceled or put on hold, while the more expensive subsea projects were drastically curtailed. Oil and gas is one of the largest industries for flow measurement, and after these projects were canceled or postponed, the companies involved bought fewer flowmeters.

The reasons for the decline in oil prices are complex, but they mainly have to do with supply and demand. Oil prices started declining in August 2014. OPEC ordinarily tries to keep prices relatively high through agreements to limit production. In November 2014, they decided not to limit production and to let the market determine prices. This resulted in an oversupply of oil, relative to demand, and oil prices plummeted.

Oil prices remained relatively low for 2 years until OPEC met again in November 2016. This time OPEC, along with Russia, agreed on production cuts. Oil prices then stabilized and began rising again in 2017. This helped the flowmeter market, but it seemed to take longer than anticipated for the positive effects to be felt in the instrumentation markets. In 2018, by contrast, oil prices were in the $60–$70 per barrel range, and the flowmeter market had a banner year. Oil prices remained relatively strong in 2019.

## OIL PRICES IN 2019 AND 2020

While oil prices were on the upswing in November and December 2019, they began slumping at the beginning of 2020. This was mainly due to the beginnings of slumping demand due to fears of the COVID-19 pandemic. The price of WTI peaked at $53.77 per barrel on February 20, 2020, and proceeded to decline from there. The continued spread of COVID-19 led to a dispute between two of the leading producers: Saudi Arabia and Russia. This dispute came to a head at OPEC's meeting in early March.

OPEC met in Vienna, along with Russia, in an Extraordinary Meeting on March 9, 2020. Russia was invited because it had participated in OPEC's oil production cuts over the past 3 years. This group (OPEC plus Russia) became known as OPEC+. At the meeting, OPEC called for a production cut of 1 million barrels per day (b/d), with Russia accounting for half of this or 500,000 b/d. Russia rejected this idea and decided it would rather engage in a price war with Saudi Arabia than join in on the cuts. Saudi Arabia retaliated by announcing its intention to increase its oil production by 1 million b/d, from 12 million to 13 million b/d. At the same time, Saudi Arabia slashed its export prices. The crude oil market responded immediately, dropping about $11 to $35 per barrel, its largest one-day drop since 1991. This move accentuated the decline in oil prices from near $50 per barrel down to the $20 per barrel range. Oil prices continued their decline in March, with WTI dipping below $20 per barrel several times. On April 6, WTI closed at $26.21 per barrel.

## Pandemic

Demand continued to decline due to the novel coronavirus pandemic, as many businesses closed and people curtailed unnecessary travel. Many restaurants began offering takeout only, and large numbers of people began working from home. This led to a devastating cut in demand for refined petroleum products. In an effort to revive slumping oil prices, OPEC met on April 9, 2020, to try to find a remedy to compensate for the sharp decline in demand.

## Another Extraordinary Meeting

In a highly anticipated Extraordinary Meeting, OPEC, led by Saudi Arabia, met with other oil-producing countries to take action on oil production quotas. The negotiations, which involved 23 countries including the G2 countries, extended over four days and resulted in the largest crude oil production cut ever agreed on. While the headline number was the 9.7 million barrels per day (b/d) that was part of the deal, non-OPEC countries contributed roughly 4 million b/d to the total. Saudi Arabia pledged an additional cut of 1.3 million b/d in May and June, and the UAE came in with a reduction of 1 million b/d. Kuwait also made a commitment to some more cuts. Besides all this, International Energy Agency (IEA) countries announced their intention to purchase oil into their reserves having the effect of taking more oil off the market. In the end, the combination of production cuts and IEA reserve purchases took from 19 to 20 million barrels per day off the market.

## The Sudden Dip

While the world oil markets responded slowly but steadily to this agreement, the initial signs were not good. One week after the OPEC+ deal was announced, the unthinkable happened. Prices of WTI plunged right through the zero value and closed at −$37.63. How is it possible for oil to have a negative price value? This means that producers had to pay storage facilities to take the oil off their hands.

The problem is that the COVID-19 pandemic dramatically reduced demand for petroleum products in the United States and the rest of the world. People weren't driving their cars as much, air travel was limited, and people were restricted from going to restaurants or in some cases from even leaving their houses. As a result, crude oil that is normally refined and then consumed was just sitting in ships or storage tanks. The problem was that the United States and the world were coming to the limit of its storage capacity for oil. Even the US national reserves in Cushing, Texas, were starting to fill up.

## SIGNS OF RECOVERY

Soon after WTI went negative, it began recovering. By early June 2020, WTI was approaching $40 per barrel, and Brent oil exceeded $40 per barrel. As economies in the United States and around the world began opening up, demand for petroleum products began coming back. At the same time, most OPEC and non-OPEC countries were abiding by the terms of the supply cuts. This had a stabilizing effect on the oil markets. After April 2020, prices continued to rise slowly through 2020 and in the beginning of 2021 prices reached the $50 per barrel range.

Oil prices in 2021 stayed above $50 per barrel until mid-February when they reached $60 per barrel. Prices stayed mostly between $60 and $65 per barrel until early June when they reached $70 per barrel. From that point on, prices stayed between $65 and $75 per barrel until early October, when they reached $80 per barrel. Prices remained high for the rest of 2021, although they dropped below $70 per barrel in December 2021.

## OIL PRICES IN 2022

In January 2022, oil prices started close to $80 per barrel, and then topped $80 per barrel on January 11. After that time, prices climbed steadily, reaching $123 per barrel on March 8. Russia's invasion of Ukraine on February 24, 2022, introduced many new uncertainties in the market. The most pronounced long-term result is likely to be a crimp in the world oil supply since some Russian oil has been taken off the market. In addition, inflation is rampant and supply chain shortages are disrupting manufacturing around the world. This is not the kind of instability that anyone wants. While instability and oil supply shortages are a recipe for even higher oil prices, most of the world is hoping for a speedy end to the hostilities.

## EFFECT ON THE FLOWMETER MARKETS

The flowmeter markets declined by about 5% in 2020. While 2021 was something of a comeback year, it was foiled in the last 2 months by the Omicron variant. In 2022, many companies are reporting higher sales as the demand for travel, dining, and entertainment return. In 2021, the worldwide flowmeter market approached the 2019 levels, but in 2022, it should exceed them.

## THE OIL AND GAS INDUSTRY AND THE OTHER PROCESS INDUSTRIES

The oil and gas industry, including refining and the petrochemical industry, is the single most important industry for the flowmeter market. Some of the most important applications include custody transfer of oil and natural gas, flue and stack gas measurement, LNG, compressed natural gas, and distribution of refined fuels (Figure 10.6). All the process industries, especially chemical and power,

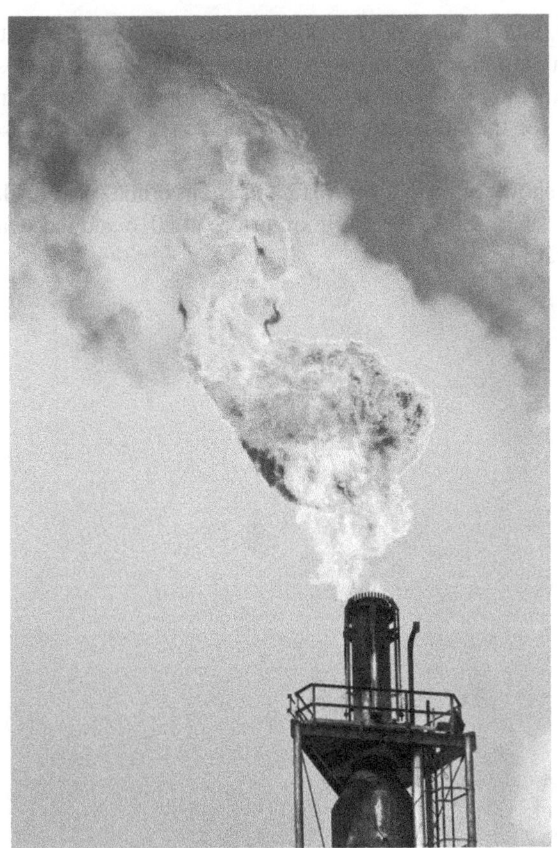

**FIGURE 10.6** Gas being flared.

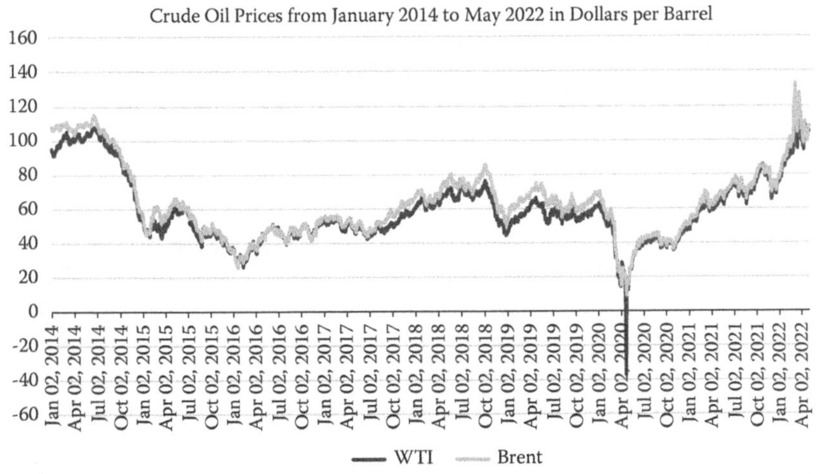


**FIGURE 10.7** Oil prices from January 2014 to May 2022.

Source: *EIA*.

rely on refined fuels as power. The most expensive and sophisticated flowmeters, especially including Coriolis, ultrasonic, differential pressure, positive displacement, and turbine, are sold into the oil and gas industry. The oil and gas industry is a barometer for the flowmeter business, and that is why it deserves so much attention.

Figure 10.7 shows oil prices for WTI and Brent crude oil from January 2014 to May 2022. This chart shows the dip in oil prices in 2016 and the steady climb in oil prices since 2020.

# 11 Four Paradoxes of Continuity

## OVERVIEW

"Every limit is a beginning as well as an ending" – George Eliot, Middlemarch. In Chapter 10, Volume I, we looked at the geometry of flow. In that chapter, I argued that points have area and that lines have width. While this view is contrary to traditional geometry, it enables us to avoid some apparent contradictions in describing the number line. It also makes possible certain facts, such as the number of points in a line, without bringing in the notion of infinity.

Another idea that runs counter to traditional geometry is that $\pi$ is only necessary because of contradictory assumptions (see Figure 11.1).

The value $r^2$ gives the geometric area of the square in the above diagram. The formula for the area of a circle, $\pi r^2$, tells us that $\pi$ squares with sides equal to radius $r$ fit into the area of a circle with radius $r$. There is a saying "You can't fit a square peg into a round hole." This saying embodies an important truth: there is no rational number of squares that will fit into a circle. This is why mathematicians have had to invent the idea of pi ($\pi$), a nonrepeating decimal with no rational value, to give an account of the area of a circle. No wonder it is called an "irrational" number.

## THE ROPE EXPERIMENT

Another curious fact about this formula has to do with the formula for the circumference of a circle. This formula is $2*\pi*r$. Without the multiplier sign (*), it is written as $2\pi r$. Now, imagine a rope or a string laid out on a table and made into a circle. Presumably, the area of this circle is $\pi r^2$ according to the traditional formula

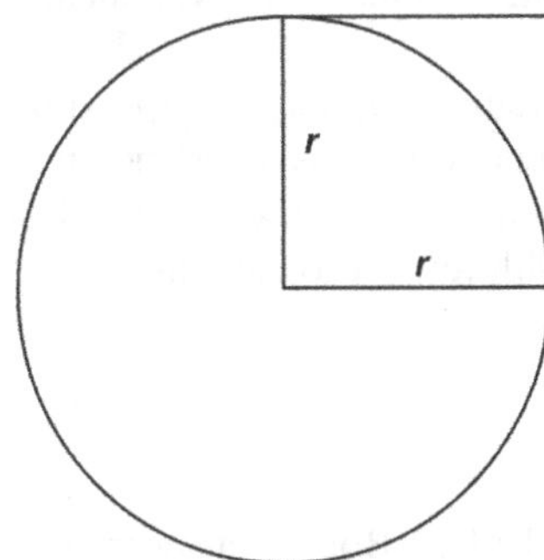


**FIGURE 11.1** The value $r^2$ as a unit for measuring the area of a circle.

DOI: 10.1201/9781003130024-11

for the area of a circle, and $r$ is the length of the radius. For simplicity's sake, let's suppose the radius is 1 inch, meaning

$$\text{Area of the circle} = \pi\ (1^2) = \pi.$$

The value of the circumference (distance around) of this circle is $2*\pi*r$

$$2 * \pi * 1 = 2 * \pi$$

More simply, we can write this without the multiplier sign (*) as $2\pi$.

Now, take the rope or string and reshape it so that it no longer forms a circle. Instead, lay it out on the table as a straight line. The length of this rope or string has not changed from the time it formed a circle to the time it formed a straight line. Yet when we measure it as a straight line, we are not going to find that it has the value $2\pi$. Instead, it will be a rational value close to 6.28 inches. There is nothing in this straight-line measurement that requires us to bring in the value of $\pi$ to account for the length of the rope or string. The reason is that it is now simply a straight line and is not in the shape of a circle. As the length of the rope or string hasn't changed between the time it was a circle and the time it took the shape of a straight line, we can only conclude that its length (circumference) when it formed a circle is also a rational value approximating 6.28, not the irrational value $2\pi$.

If this is true, we need to take another look at the formula for the area of a circle. The problem is that the current formula is with the unit of measurement. As long as we use a square as the unit of measurement for the area of a circle, there will be no way to give the exact value of the area of the circle. Only if we can find a different unit of measurement to measure the area of circles can we arrive at a formula that gives the area of a circle. The problem is that the areas of squares and circles are incommensurable, meaning that there is no common unit that can be used to render the area of both squares and circles. Or at least if there is such a unit, it hasn't yet been proposed or discovered.

I suggested using a round inch as a unit of measurement for a circular area. While this will give us rational values for the circular areas, it won't give us rational values for squares. Just as you can't fit a square peg into a round hole, so you can't fit a round peg into a square hole, unless it's significantly smaller. And even then, it won't take up the entire space of the square.

Another approach is to see if there is some hybrid figure, possibly incorporating curved or circular areas and also straight lines, which could serve as a common unit of measurement. If there is such a figure, I don't know what it is. However, this is a question worth exploring. A similar problem exists with respect to calculus, as we shall see in the next section.

## A FLAW IN CALCULUS

Calculus has similar defects. What is wrong with calculus? Bob A. answers this question on the forum "Wyzant Ask an Expert" as clearly as any I've seen:

> In calculus you find the area under a curve by taking the integral.
>
> To find the area between two specific places you evaluate the definite integral between those limits.
>
> There are numerical ways to find the area or to evaluate an integral.
>
> You split the region under the curve between the two limits into finite slices (columns).
>
> There are many different ways you can do this.
>
> Two common ways to do this are with a rectangular slice or a trapezoidal slice.

For the rectangular method, you can have the top left of the rectangle be a point on the curve, have the top right be a point on the curve, or the more accurate way is to have the center point of the top of the rectangle on the curve. That way there will be a little of the rectangle above the curve and a little below coming to an average of about right. Then, you calculate the area of all these little, tall rectangles and add them up. On a computer this is easy, the computer can reduce the widths of the rectangles smaller and smaller → approaching a limit of zero or some value of precision you want.

The concept of limit is fundamental to calculus. In the above account, the number of rectangles drawn under a curve "approaches a limit of zero or some value of precision you want." Of course, if the widths of the rectangles are zero, then they are like a rectangle with no area. But a rectangle with no width is no rectangle at all. It's simply a line with no width, like Euclid's "breadthless area."

Some people say that it is possible to describe calculus without relying on the notion of infinity. In a video description of calculus called "Introduction to Integral Calculus" by the Khan Academy, the narrator describes the same process that is described in the above quote, of drawing a series of rectangles below the curve, each of which approximates the area below the curve. He then says

> As long as we're using a finite number, we can always have the [rectangle widths] smaller, and we have more of these rectangles … What happens as [the rectangles] get thinner and thinner and thinner, [the number of rectangles] n gets larger and larger and larger, the [rectangle widths] get infinitesimally small and n (the number of rectangles) approaches infinity. You're probably sensing something; that maybe we could think of the limit as n [the number of rectangles] approaches infinity. This notion of getting better and better approximations as n approaches infinity is the core idea of integral calculus.

Calculus is another example where we start with a flawed assumption, then have to bring in the concept of infinity to bridge the gap between the results of our assumptions and mathematical truth. The assumption of calculus is that it's possible to accurately represent the area under a curve by drawing infinitely many rectangles that become so small that their width is zero and then call the limit of that process the value of the area under a curve. The problem with this process is that a rectangle with zero width is not a rectangle, and it is not possible to complete an infinite process.

Some people say, "But calculus got us to the moon." Perhaps it did because calculus gets us close enough to the actual value of the area under a curve to be useful for mechanical calculations. But this doesn't mean it gives us the exact value under the curve; it only gives an approximation.

This is similar to analyzing the area of a circle in terms of $\pi$. By starting with the flawed assumption that you can provide the area of a circle by determining how many squares will fit into it, you are forced to end up with an irrational number as the value. The result is that $\pi r^2$ doesn't give an exact value for the area of a circle; it only gives an approximation.

It's all about the assumptions we start with. The assumption that points have no area and that there are infinitely many points on a line leads to Zeno's Paradox, as I discuss in Chapter 5 of my book *Tao of Measurement* (Table 11.1).

The way to solve the above paradoxes is to develop a geometry that recognizes that points have area and lines have width. In addition, it is worth developing a geometry with a circular or curved unit of measurement that can give us a rational

**TABLE 11.1**
**Four Paradoxes of Continuity**

| Flawed Assumption | Compensating Flawed Conclusion | Solution |
|---|---|---|
| The way to determine the area of a circle is to find out how many squares it contains. | The area of a circle is an irrational value containing $\pi$, an irrational nonrepeating number. | Use a circular or curved area instead of a square as the unit of measurement for circular area. |
| The way to determine the area under a curve is by drawing a series of rectangles that approximate the area. | Imagine that this series of rectangles is infinite in number and gets ever smaller until one has zero width. | Use a different unit of measurement for the area under a curve – possibly a curved area. |
| The number line is made up of discrete points with no area. | There are infinitely many points with no area in a line and another point can always be put between any two points. | Points line on the line, not in the line. Specify the unit of measurement ahead of time. There are only finitely many points on the line and points are not dimensionless; points have an area. |
| Zeno's Paradox: To go from Point A to Point B, someone must first go halfway there, then halfway again, and halfway again. | There is no end to this series because an infinite number of steps is required, and an infinite series cannot be completed. | When someone goes halfway between Point A and Point B, he or she is never at the halfway point because they only pass through or over that point. The person is only at the halfway point if he or she stops at that point. If the person keeps going without stopping, he or she will reach Point B without having to go through an infinite number of steps. |

value for the area of a circle and that can give us rational values for the area under a curve. This proposed unit of measurement would replace the current square or rectangle that cannot give us the exact value for the area of a circle or the area under a curve.

It is time to take a new look at Euclid's definitions, postulates, and axioms, which have governed mathematics for over 2,000 years. In Euclid's Book I, Definition 2, Euclid says "A line is a breadthless length." He does not define what he means by breadth, but what he is saying here is that a line has one dimension only and it does not have width. This is the root of some of the paradoxes presented in Table 11.1.

It is worth pointing out that the non-Euclidean geometries that have been developed in the past two centuries do not address the issue of whether a line has width, or a point has an area. Instead, they focus on Euclid's fifth postulate regarding parallel lines. It is commendable that these geometries have been developed but they do not address the issues I am raising here.

In my book *Tao of Measurement,* I propose 12 new axioms that could be the basis for a new circular geometry. Considering all the effort that has been put into analyzing the value of $\pi$ and exploring the meaning of irrational numbers, to say nothing of the time spent on calculus, I think it would be worthwhile to explore an alternative geometry that doesn't require irrational numbers. I know that set theorists and logicians, such as Leslie Tharp and the legendary Hao Wang, with whom I studied set theory and logic at Rockefeller University, have no problem with irrational numbers or infinite sets. As I have great respect for their intellects, I have to believe that set theory is a valid enterprise, even though it does treat an infinite series of numbers as if it can be a completed set. However, I believed then and I still believe today that bringing in the idea of infinity is simply a way to compensate for flawed assumptions. I would like to see anyone present a definition of infinite that goes beyond "not finite." In reality, an infinite series can never be completed and no set of numbers that is infinite can exist as a completed set. While set theory as it exists today has its value, I would like to see some effort put into exploring alternative geometries that do not rely on the notion of infinity.

Most of all, we need a new unit of measurement that allows us to give rational values to the areas of both circles and squares, as well as the area under a curve. The alternative to this is to develop a circular geometry with a circle as the unit of measurement. This will enable us to assign rational values to circular areas. The downside is that the areas of squares will most likely be given in irrational values. The result of this approach is that we have two geometries: one for straight-edged figures and one for circular or curved figures. If this is a mathematical reality, then this is something we cannot change. However, if we can find a unit of measurement that is common to straight-line figures and circular figures, we can have a single geometry that gives rational values for all these figures.

# 12 Sensing and Measuring

## THEORY OF SENSING

So far in this book, we have examined multiple types of flowmeters and the way they measure flow. Chapter 3 covers some basic concepts of flow. Chapter 3 defines conventional flowmeters and how they differ from new-technology flowmeters. Chapters 4–9 discuss the different types of conventional flowmeter technologies, including the following:

Chapter 4 – Differential Pressure Transmitters
Chapter 5 – Primary Elements
Chapter 6 – Positive Displacement Flowmeters
Chapter 7 – Turbine Flowmeters
Chapter 8 – Open Channel Flowmeters
Chapter 9 – Variable Area Flowmeters
Chapter 10 – Applications
Chapter 11 – Four Paradoxes of Continuity
Chapter 12 – Sensing and Measuring

In the earlier chapters, we considered flowmeters as a whole unit, except that we separated differential pressure (DP) transmitters from primary elements. What we did not do so far is examine how a flowmeter is structured. There is a common structure to all the flowmeters we have considered. In this chapter, we look at the structure that is common to most flowmeters.

Most flowmeters are structured as follows:

Sensor + Transmitter = Sensing and Measuring Flow

In most flowmeters, a sensor senses a physical variable. Coriolis meters sense a difference in the phase shift of vibrating tubes upstream vs. downstream, and ultrasonic meters sense sound waves transmitted across the flow. Thermal flowmeters sense a difference in heat upstream and downstream. DP transmitters sense a difference between the pressure upstream and downstream of a primary element. Turbine meters rotate in proportion to flow, and in this way sense the speed of the flow.

It is important to distinguish between sensing and measuring, as flowmeters do both. However, it's more complicated than that, because there is usually a transducer involved that takes the signal from the sensor, amplifies it, and passes it on to the transmitter. The transmitter does the measuring. But before we get too deeply involved in what is a complex process, let's begin by looking at what a sensor is, and how it differs from measuring.

DOI: 10.1201/9781003130024-12

## Sensing and Measuring

One important distinction that underlies the discussion in this book is the distinction between sensing and measuring. Flowmeters may not sense flow directly, but they all contain sensors that sense some parameter that is associated with flowrate. For example, magnetic flowmeters have electrodes that sense the voltage created when a conductive liquid passes through a magnetic field and uses this information to compute flowrate. Coriolis flowmeters sense the impact of fluid momentum on a vibrating tube and use this information to determine mass flow.

Flowmeters contain sensors, but they are also measuring devices. A measuring device contains a unit of measurement, and this unit of measurement is used to determine a quantity that determines how much of that unit something has. For example, if a flowmeter has gallons per minute as a unit of measurement, it uses information from the sensor to determine how many gallons of fluid are passing through a pipe in a given period of time. Just as thermometers have Kelvin, Celsius, and Fahrenheit scales, so flowmeters have different units in which they measure. Some units are Imperial (American) and others are metric (SI). The units can often be programmed into the flowmeter.

## What Is a Sensor?

In order for something to be a sensor, it must respond in a predictable way to some property or parameter in the object it is sensing. This means there must be some interaction between the object being sensed and the sensor. However, not all forms of interaction count as sensing. A stick of wood placed into a river isn't sensing the flow; it just deflects it. A bar of steel does not sense light even if it reflects it because the reflected light is not used to determine the presence of a quality in the bar of steel; the light is simply reflected.

Sensors that are sensing a physical quality or property in an object share the following characteristics:

1. They are made from some kind of physical material.
2. This material is sensitive to changes in the qualities of a physical property of the object.
3. The sensing material or property responds to changes in the qualities of a physical property in the object in a predictable way.
4. A converter, transducer, or amplifier measures these changes, amplifies them, and passes them to another device such as a transmitter or recorder.

What kind of relationship has to exist between a sensor and the property of the sensed object? It has to be a predictable relationship, at the very least. A mercury thermometer whose readings vary wildly and inconsistently with the temperature is not sensing it, even if it is responding to it. An outside thermometer that reads 70° when it is 30° outside and 86° when it is 40° is not sensing the temperature but instead responding in a seemingly arbitrary way to it. The idea of sensing contains the idea of truth, so that a sensor portrays an objective value, within certain bounds of correctness. Even so, a thermometer that is a few degrees off is still sensing the temperature even if it is not completely accurate.

What is the relation between a sensor and the property or a sensed object that makes it a sensor? Because of the element of truth or accuracy that is implied in saying that something is a sensor, they must be related in a predictable way. This means that the sensor is following a rule in the presence of the sensed object or property. This rule may not always be known, but it must exist. It is the rule that formulates the predictable relation between the sensor and the property or sensed object. In the case of a mercury thermometer, when the mercury is at a certain height, it reads 50°, and when it is at a different height it reads 80°. The height of the mercury in response to temperature depends on the expansion powers of mercury.

This enables us to formulate a fifth principle relating to sensors:

5. When a sensor exists and senses the presence of a property or an object, it is following a rule that formulates the relation between the sensor and the sensed object or property. This rule may or may not have been explicitly formulated.

Some sensors do not become parts of flowmeters. Pressure and temperature sensors read the pressure and temperature and may display it but may not be part of a flowmeter. However, some flowmeters such as multivariable flowmeters incorporate temperature and pressure sensors and use this information to compute flowrate. A flow sensor typically senses a variable that is associated with computing flowrate. For example, in a vortex meter, ultrasonic sensors are often used to count the number of vortices generated by the bluff body. The number of vortices generated is used to compute flowrate.

### The Amplifier or Transducer

The signal from a sensor is usually pretty weak. In order for it to be used by a transmitter, the signal must be amplified. This is done by an amplifier or transducer, which amplifies and conditions the signal so it is stronger than the output from the sensor. The output from the sensor serves as input to the amplifier or transducer. The output from the conditioner or transducer serves as input to the transmitter, which is programmed to calculate flowrate based on the sensor's input.

### Sensor and Transmitter

Some flowmeters are sold as integrated devices, with the sensor and transmitter integrated together into a flowmeter. For other flowmeters, the sensor is sold separately from the transmitter. This is generally true for magnetic flowmeters. The sensor component consists of a tube containing magnetic coils, with electrodes mounted on the either side of the flowtube. The magnetic coils create a magnetic field when conductive liquid moves through the tube. The electrodes sense the strength of the voltage of the magnetic field. This information is sent to the transmitter. As flowrate is proportional to the strength of the voltage generated by the motion of the conductive medium, the transmitter computes flowrate based on input from the electrodes, along with other information.

## WHAT IS MEASUREMENT, AND HOW DOES IT DIFFER FROM SENSING?

Sensing and measuring are closely connected, and it is important to understand both, especially since flowmeters both sense and measure. As described above, sensors sense variations in some physical parameters and emit a signal. This signal is amplified by a converter or transducer and sent to a transmitter, which accepts it as input.

Measuring does not involve or include sensing, although a sensor can be an input to a measuring device. Measuring requires a unit of measurement, and measurement determines how many or much of that unit of measurement something has. For example, if our measuring device is a yardstick, the unit of measurement is in inches. Yardsticks typically measure length or height. A yardstick has a series of marks on the side to mark off inches. We use a yardstick by comparing these marks to an object we want to measure the length or height of. The length of a table is determined by comparing the markings on the yardstick to the end point of the table. So, if the table is 20-inch long, the end of the table is equal to the 20-inch mark on the yardstick.

A yardstick does not sense length or height. It does not have a sensor that responds to some parameter in an object and creates an output that represents that parameter. Instead, a yardstick is simply a device for measuring how many units in terms of inches or feet something has. Likewise, a measuring cup measures the volume and is often used in baking. If a recipe requires two cups of milk, we can use the measuring cup by filling it to the top line twice with milk. This is somewhat like the way a positive displacement flowmeter works.

Units of measurement differ in different parts of the world. In the United States, most people still use the inch, foot, yard, and mile as the units of length. These are called the imperial units. Most European countries use the metric system, which is based on the meter and the kilometer. There are correspondingly different imperial and metric units for area, volume, and weight or mass.

The different systems of measurement, how they were derived, and how they compare to each other is a subject for a book in itself. I discuss this topic in more detail in Chapter Eight of my book *Tao of Measurement*. It's a fascinating story. What is important here is to consider how flowmeters measure flow.

The units of measurement used by flowmeters differ with the type of fluid being measured. For liquids, a standard unit of measurement is gallons or liters. Cubic centimeters are also used as a unit of measurement for liquids. A cubic centimeter is the volume of a cube that has 1 centimeter on each side. A cubic centimeter is equal to 1/100th of a liter. Gases are generally measured in cubic feet ($ft^3$) or cubic meters ($m^3$).

We often talk about the quantity of flow in terms of the amount of flow. For example, we might say, "Six gallons of water flowed into the tub." When we add time, we are talking about flowrate. For example, we might say that a test stand is flowing at the rate of 35 gallons per minute (gpm). Once we know the flowrate and the time, we know the quantity of flow. For example, if a test stand flows 35 gpm for 5 minutes, we know that the volume that has flowed through the test stand is 175 gallons.

In these two volumes, we have examined 10 types of flowmeters. While they vary by technology, what most of them have in common is that they compute flow by using flowrate and time, and then display it. They also may send it to a distributed control system (DCS), a programmable logic controller (PLC), or output it in the form of a communication protocol to another instrument or measuring or controlling device. The main exceptions to this are positive displacement meters, which simply capture the flow in a known amount and then count how many times this is done to get a total volume. Variable area meters are another example because many of them need to be read manually and they do not compute flowrate on their own. However, some manufacturers are now putting transmitters on their variable area flowmeters, so that these meters can both display a flow value and send a signal to a recorder, controller, instrument, or other device.

Communication protocols have become increasingly important. After "smart" flowmeters were introduced by Honeywell in 1983, it took about 10 years for the idea to "catch on" in the field. Eventually, the Fieldbus Foundation was formed and the Profibus protocol emerged. The most popular communication protocol was HART, which was a digital signal interposed on top of a 4–20 mA output. In recent years, new protocols have emerged such as Modbus, Devicenet, and others. Some companies continue to use their own proprietary protocols, such as Yokogawa's Brain, but this trend is diminishing due to the need for interoperability among devices.

## Mechanical, Electronic, and Biological Sensors

While there are many types of sensors, the following three types are considered in this section:

- Mechanical
- Electronic
- Biological

Most sensors were mechanical before the advent of electricity and electronics. Mercury in a glass thermometer that measures temperature is one example. Another is a Bourdon tube, which measures pressure. A pressure gauge is a relatively simple mechanical sensor that displays a pressure reading on a dial. Other types of mechanical sensors include force sensors, accelerometers, and gyroscopes.

## Electronic Sensors

Most flowmeters today use electronic sensors, although some use a combination of mechanical and electronic sensors. Positive displacement meters physically capture the flow in small compartments, then count how often this is done. This is a way to measure volumetric flow. The physical capturing of the flow is the mechanical part, while the counting is done electronically. This information is fed to a transmitter, also called a register. Some of these meters have odometer-like devices that display the volumetric flow in numbers. Other more recent models contain transmitters that display the readings electronically.

A turbine meter can be viewed as a combination of mechanical and electronic sensors. The rotor that spins in proportion to flowrate is mechanical, but the magnetic pick-up coil that counts the revolutions is electronic. Furthermore, in most turbine meters, the data from the pick-up coil is related to an electronic transmitter that computes flowrate. It may then transmit a signal in the form of a 4–20 mA signal, an analog signal or a digital signal. The digital signal may incorporate communication protocols such as Foundation Fieldbus, Profibus, Modbus, or some other protocol. Alternatively, the output from the turbine flowmeter may go to a recorder or to a DCS.

Most other types of flowmeters use electronic sensors. A Coriolis meter detects the phase differences in a vibrating tube that responds to the momentum of the fluid. Thermal flowmeters use temperature sensors to detect temperature changes in a flowstream to which heat has been added. DP flowmeters measure the difference in pressure upstream and downstream of a primary element and use this pressure difference to compute flow, using Bernoulli's theorem.

There is more detail in the individual Chapters 4–9 that explain how these different flowmeters work. In some cases, the principle of operation is shown with an illustration.

## Biological Sensors

Human beings traditionally are viewed as having five sense organs: eyes, ears, tongue (taste), nose (smell), and skin (touch). The way in which humans process sensory information is somewhat like the way in which electronic flowmeters operate. However, the human sensory system, brain, and mind is far more complex than any flowmeter. There is also much about the human brain and mind that we don't understand. Even so, considering vision, when my eyes see an object, an image of that object is transformed into biological patterns in both eyes. The two patterns are then merged by the brain into a single image; otherwise, we would be "seeing double" all the time. How the image is processed by the brain and then becomes a part of consciousness is something we still do not clearly understand.

One way that our senses work like electronic sensors is that they both seem to follow rules. When I look at a blue object, or an object of a particular shade, I will have the same experience each time. That is, of course, if my eyes are working properly. This element of predictability is fundamental to our experience. Honey and sugar taste sweet; the taste only varies with the type of honey or sugar. The sound of a train horn is the same, provided it is the same horn. Obviously, horn sounds vary with different types of horns

One of the greatest philosophers who ever lived, Descartes, did us a disservice by setting up mind and body as incompatible entities. According to Descartes, bodies take up space and are physical, while minds are immaterial. This makes it very difficult to explain how they can interact.

Another way of viewing the situation is that the mind has both physical and mental aspects. When I have a thought that I want to pick up a cup, the physical aspect of the thought initiates a sequence of events in my brain that direct my muscles to pick up the cup. This sequence of events is set up through learned experience, much like when children learn to talk. This shows how a thought can have a direct impact on the body.

How does the body interact with the mind? If I have a cut on my hand, sensory impulses are sent to the brain reporting this event. This initiates a series of neural impulses in my brain that are associated with a feeling of pain. The feeling of pain has both neural and mental aspects. Viewing mind–body interaction in this way removes the dilemma introduced by Descartes' mind/body dichotomy and makes it possible for thoughts (mind) to influence the body and for the body (physical cut) to interact with the mind.

## FRONTIERS OF RESEARCH

### The Measurement of Time

It is difficult to separate time from measurement of time, but time existed long before it was measured. Scientists put the age of the universe at about 14 billion years, and the age of the earth at 4.5 billion years. Our earliest ancestors appeared about six million years ago. What we call *Homo sapiens* only came into being in the last 300,000 years. These early human beings were not wearing watches, and mechanical clocks were not invented until the 14th century. Yet time existed long before that, including before the earth began. In *Tao of Measurement*, I discuss whether time could have existed before our universe began, and what to make of the phrase "the beginning of time."

Time measurement is a human construct that we create to enable us to navigate our lives and it relies on both natural and mechanical phenomena. In ancient times, humans marked the passage of time by observing the sun, moon, and stars, and the passing of the seasons. Once clocks were invented in the 14th century, people began relying more on these mechanical devices to take account of time. In the 20th century, time was perceived more as being continuous since most clocks had sweeping second hands. Today, digital clocks make time seem more like a series of unconnected points, somewhat like the idea of a number line composed of infinitely many discrete points.

### Do Clocks Sense Time?

A clock does not sense time in the way in which a temperature sensor senses temperature. Instead, a clock or watch runs parallel with time. A mechanical clock is entirely self-directed and responds to a predetermined movement or a physical setting. A digital clock has a gearing mechanism or a crystal oscillator and displays time in the form of numbers. A temperature sensor, by contrast, responds with a physical change to heat or cold, which is in the form of molecular motion. It is usually attached to a transducer that amplifies the signal and interprets the response of the sensor as indicating a specific temperature. It may also pass the signal to a transmitter, or to a display that shows the temperature in degrees. A simpler example is a thermometer that contains mercury. The mercury rises and falls in response to the temperature. While the thermometer is also a measuring device, due to the scale on one or both sides of the mercury, it indicates the temperature by responding to heat and cold in a predictable way.

## Decimal Time

In *Tao of Measurement*, I trace the development of clocks from Huygen's invention of the pendulum clock all the way to the atomic clock of today. What is especially interesting here is the development of the units that measure time. There are different systems of decimal time, but what they have in common is an attempt to change our time from one based on 24 hour days to one derived from a base 10 system. In the preceding paragraphs, we discussed the three different sources for our current system of measuring time. This necessarily introduces some conflicts and inconsistencies in the system due to the different origins of the time systems.

Some decimal systems divide the day into 10 hours rather than 24 hours. These systems typically allocate 100 minutes to each hour and 100 seconds to each minute. Decimal time has never been in effect for a broad population except for a brief period just after the French Revolution. The French instituted decimal time on November 24, 1783. French Revolutionary time had a 10-hour day, 100-minute hours, and 100-second minutes. Clocks were manufactured to reflect the new time (Figure 12.1). However, people found it difficult to make the transition to this new system, part of which involved replacing every clock and watch in the country. The new time system was abandoned after 17 months.

## Advantages of Decimal Time

Decimal time does have some advantages over our sundial-based system of 24 hour days. These advantages mainly derive from the fact that most of our calculations are done in a base 10 system. When using a base 10 system to compute values in a 24 hour system, it works out, but the values are not very intuitive. For example, 60% of 24 hours is 14.4 hours. This works out to 14 hours and 24 minutes. In decimal time, 60% of 10 hours is 6 hours, which works out to 600 minutes.

Another perhaps more familiar example is in payroll systems. Because most employees are used to our base 10 system of calculating, it is natural to report hours in base 10 format. For example, if someone reports that they worked 16.5 hours 1 week, this amounts to 16 hours and 30 minutes. Working 16.75 hours amounts to 16 hours and 45 minutes. In a decimal system, 16.5 hours would be 16 hours and 50 minutes, while 16.75 hours would be 16 hours and 75 minutes.

## Flowtime

I have proposed a different system of decimal time that preserves some of the French methods of decimal time while keeping some of our traditional ways of

**FIGURE 12.1** The face of a decimal clock in neo-classical style made by Pierre Daniel Destigny in Rouen, France, between 1798 and 1805. This clock is part of the collection of the Fitzwilliam Museum, Cambridge — Photo courtesy of DeFacto [CC-BY-SA 4.0] https://commons.wikimedia.org/wiki/File:Decimal_Clock_face_by_Pierre_Daniel_Destigny_1798–1805.jpg.

measuring time. The reason it is called "flowtime" is that it more closely approximates continuity by introducing more discrete points within hours and minutes while preserving the 24-hour clock. I think that changing today from a 24-hour clock, or two 12-hour periods to mark the day and night, would be too difficult for people to get used to. No doubt this is one of the reasons that the French experiment with decimal time failed (Figure 12.2).

The proposal of flowtime is to keep the number of hours in the day as 24 but to change the number of minutes in an hour from 60 to 100. Likewise, in flowtime, there are 100 seconds in a minute instead of the usual 60. This preserves our way of writing hours but changes our way of writing minutes and seconds. As a result, 4:30 in conventional time becomes 4:50, meaning four hours and 50 minutes. Likewise, 4:45 becomes 4:75, or four hours and 75 minutes. One way to distinguish conventional time from flowtime is to drop the colon that is used to denote conventional time and replace it with a period to denote flowtime. So, 4:50 can be written as 4.50, and 4:75 can be written as 4.75.

Converting between conventional time is easy. Simply take the minutes or seconds in traditional time and multiply by 5/3, or 1 2/3. The result is the number of minutes or seconds in flowtime. The hour remains the same. For example, 4:30 in traditional time is 30 × 5/3 = 50. So, 4:30 in traditional time is 4.50 in flowtime.

**FIGURE 12.2** Flowtime clock displays both flowtime and regular time—Photo courtesy of Jesse Yoder.

The value 4:30:15, meaning 4 hours, 30 minutes, and 15 seconds comes out to 4.50.25 in flowtime, meaning 4 hours, 50 minutes, and 25 seconds.

In the examples I've picked, the numbers come out to round numbers when converted. Unfortunately, this is not always the case. For example, 40 in traditional minutes comes out to 66.66 when multiplied by 5/3. In this case, the number can be rounded up to 67, so that 4:40 in traditional time is 4.67 in flowtime. However, this is not surprising since 40 is 2/3 of 60 and representing 2/3 in decimal form always comes out to the rational but repeating .6666 or .67. This is a problem that naturally results from two fundamentally incompatible systems. Minutes and seconds are calculated based on 60, while in flowtime, the base value is 100. Even so, flowtime will still have similar difficulties in rendering 2/3 of an hour, since that will still come out to 66.66 or 67 minutes. There is a clock that constantly displays flowtime 24/7 at the bottom of the website https://www.flowresearch.com.

## Why Change to Flowtime?

1. Flowtime divides time into smaller quantities. The number of minutes per hour changes from 60 to 100. As of now, there are 1,440 minutes per day, while in flowtime there are 2,400 minutes per day. Now there are 3,600 seconds in 1 hour; in flowtime, there are 10,000 seconds per hour. Having smaller units of time makes it possible to break tasks down into more discrete periods.

2. Flowtime is more consistent with digital clocks. Under traditional time, counting down from 1 minute 30 seconds to 59 seconds introduces a gap at the 1-minute mark. Instead of going to 99 seconds, under traditional time, we count down from 1 minute to 59 seconds. This introduces an unintuitive gap as the time reaches the 1-minute mark. Under flowtime, the counting would continue to 99, 98, and then down to zero.
3. Flowtime may be useful for some time-driven sporting events where split-second timing is required. This is already done in the National Basketball Association (NBA). As the clock winds down, the seconds get divided into ten equal segments. This is actually carrying flowtime one step further by further dividing seconds into even smaller units of 1/10th of a second.
4. There may be some applications for computers that already have had to create much smaller units of time to measure the time that computers use in their operations. Computers already use nanoseconds to measure the speed of logic and memory chips. A nanosecond is one billionth of a second. Electricity, far from being instantaneous, travels about one foot in a wire in a nanosecond. It could be that flowtime could make computer calculations easier because it is based on our traditional decimal base 10 system.
5. There may be other areas of life where there is an advantage to having smaller units of measurement. If computers need to operate in nanoseconds, there may be other electronic devices that could benefit from a system that already has more discrete points of measurement. Possible areas are semiconductor manufacturing, telecommunications, pulsed lasers, and related aspects of electronics. Possibly flowtime could be used in the transmitters of flowmeters and possibly in communication protocols to reflect and quantify faster speeds of operation.

It is unlikely that society as a whole would ever switch to flowtime, despite all its advantages. Doing this today would require replacing billions of clocks, billions of watches, and reprogramming billions of computers. For example, it is estimated that there are 5 billion clocks in the world. Add to this the number of watches, and the number of computers and you have a truly gigantic task on your hands. This would make the Year 2000 scare when many people thought that a large number of computers would stop working and have to be reprogrammed, look like a minor operation.

Think of flowtime as being like cars. There are small economy cars that provide basic transportation and are relatively low cost. There are more expensive cars that are better built, cost more, but are more reliable. Then, there are luxury cars that people drive partly for show, although many of these cars are also very well built and long-lasting. Finally, there are race cars that are small, incredibly fast, and designed for racing.

No one would propose that everyone should drive the same kind of car regardless of cost or need. Of course, there are advantages to everyone having the same time system. Flowtime is like a race car that could work effectively for certain applications where there is an advantage to having smaller units of time measurement. The existence of nanoseconds already shows that such applications exist. And as scientists measure the speed of light and look far into the earliest moments in time, possibly flowtime could play a role in those calculations.

# Index